T0231415

FOOD LOSS AND
WASTE REDUCTION

Technical Solutions for Cleaner Production

FOOD LOSS AND WASTE REDUCTION

Technical Solutions for Cleaner Production

Edited by
Laxmikant S. Badwaik, PhD
Cristóbal Noé Aguilar, PhD
A. K. Haghi, PhD

First edition published 2022

Apple Academic Press Inc.
1265 Goldenrod Circle, NE,
Palm Bay, FL 32905 USA

4164 Lakeshore Road, Burlington,
ON, L7L 1A4 Canada

CRC Press
6000 Broken Sound Parkway NW,
Suite 300, Boca Raton, FL 33487-2742 USA

2 Park Square, Milton Park,
Abingdon, Oxon, OX14 4RN UK

© 2022 Apple Academic Press, Inc.

Apple Academic Press exclusively co-publishes with CRC Press, an imprint of Taylor & Francis Group, LLC

Library and Archives Canada Cataloguing in Publication

Title: Food loss and waste reduction : technical solutions for cleaner production / edited by Laxmikant S. Badwaik, PhD, Cristóbal Noé Aguilar, PhD, A.K. Haghi, PhD.

Names: Badwaik, Laxmikant S., editor. | Aguilar, Cristóbal Noé, editor. | Haghi, A. K., editor.

Description: First edition. | Includes bibliographical references and index.

Identifiers: Canadiana (print) 20210168692 | Canadiana (ebook) 20210168951 | ISBN 9781771889391 (hardcover) | ISBN 9781774638217 (softcover) | ISBN 9781003083900 (ebook)

Subjects: LCSH: Food industry and trade—Waste minimization. | LCSH: Agricultural wastes—Management. | LCSH: Food waste. | LCSH: Food waste—Prevention.

Classification: LCC TD899.F585 F63 2022 | DDC 664.0028/6—dc23

Library of Congress Cataloging-in-Publication Data

Names: Badwaik, Laxmikant S., editor. | Aguilar, Cristóbal Noé, editor. | Haghi, A. K., editor.

Title: Food loss and waste reduction : technical solutions for cleaner production / edited by Laxmikant S. Badwaik, Cristobal Noé Aguilar, A.K. Haghi.

Description: Frist edition. | Palm Bay, FL : Apple Academic Press, 2021. | Includes bibliographical references and index. | Summary: "Focusing on the crucial sustainability challenge of reducing food loss at the level of consumer society, this volume provides an in-depth, research-based overview of this multifaceted problem. Food Loss and Waste Reduction: Technical Solutions for Cleaner Production considers the myriad environmental, economic, social, and ethical factors associated with enormous amount of food waste, which also ends up wasting water, air, electricity, and fuel, which are necessary for food processing. This book uniquely examines the social and cultural views of food waste management, shedding new light on the topic by emphasizing the consumer/household perspective throughout. Drawing on a wide variety of disciplines, the book presents philosophical reflections, practical examples and case studies, and potential solutions to the problem of increasing food waste. The special feature of the book is that it covers different developments made right from the basic technologies generated for waste management to recent advancements and future areas of research. Topics include Edible/biodegradable hydrogels as a packaging material Using encapsulation as a food preservation technique Effective utilization of food waste Cold plasma in food processing Food waste in transportation Use of various forms of energy for food processing Recovery, separation, and purification of biocompounds with potential application in the food industry Effective use of cereal and wine by-products This book will be of great interest to postgraduate students and researchers in environmental policy, waste management, social marketing ,and consumer behavior, as well as policymakers and practitioners in consumer issues and business"-- Provided by publisher.

Identifiers: LCCN 2021012457 (print) | LCCN 2021012458 (ebook) | ISBN 9781771889391 (hardback) | ISBN 9781774638217 (paperback) | ISBN 9781003083900 (ebook)

Subjects: LCSH: Food industry and trade--Waste minimization. | Agricultural wastes--Management. | Food waste.

Classification: LCC TD899.F585 F665 2021 (print) | LCC TD899.F585 (ebook) | DDC 664.0028/6--dc23

LC record available at https://lccn.loc.gov/2021012457
LC ebook record available at https://lccn.loc.gov/2021012458

ISBN: 978-1-77188-939-1 (hbk)
ISBN: 978-1-77463-821-7 (pbk)
ISBN: 978-1-00308-390-0 (ebk)

About the Editors

Laxmikant S. Badwaik, PhD

Associate Professor, Department of Food Engineering and Technology, Tezpur University, Tezpur, Assam, India

Laxmikant S. Badwaik, PhD, is Associate Professor of the Department of Food Engineering and Technology at Tezpur University, Assam, India. He is also serving as Deputy Director of the National Accreditation Board for Testing and Calibration Laboratories-accredited Food Quality Control Laboratory of Tezpur University, Tezpur. He has been involved in teaching and research for the last 15 years in the area of food engineering and technology. His main areas of research are food quality and safety, food packaging, and food processing waste utilization. He has published more than 25 research articles and chapters in various journals and books and also edited one book, *Innovations in Food Processing Technology.* He has handled various funded research projects related to food processing and food packaging. He holds B Tech in Food Technology from the University Department of Chemical Technology (UDCT), S.G.B. Amravati University, Amravati, Maharashtra, India; M Tech in Food Engineering and Technology from the Sant Longowal Institute of Engineering and Technology (SLIET), Longowal, Punjab, India; and a PhD in Food Engineering and Technology from Tezpur University, India.

laxmikantbadwaik@gmail.com

Cristóbal Noé Aguilar, PhD

Director of Research and Postgraduate Programs at Universidad Autonoma de Coahuila, Mexico

Cristóbal Noé Aguilar, PhD, is Director of Research and Postgraduate Programs at the Universidad Autonoma de Coahuila, Mexico. Dr. Aguilar has published more than 160 papers in indexed journals, more than 40 articles in Mexican journals, and 250 contributions in scientific meetings. He has also published many book chapters, several Mexican books, four editions of international books, and more. He has been awarded several prizes and awards, the most important of which are the National Prize of Research 2010 from the Mexican Academy of Sciences; the Prize "Carlos Casas Campillo 2008" from the Mexican Society of Biotechnology and Bioengineering; National Prize AgroBio—2005; and the Mexican Prize in Food Science and Technology. Dr. Aguilar is a member of the Mexican Academy of Science, the International Bioprocessing Association, Mexican Academy of Sciences, Mexican Society for Biotechnology and Bioengineering, and the Mexican Association for Food Science and Biotechnology. He has developed more than 21 research projects, including six international exchange projects. His PhD in Fermentation Biotechnology was awarded by the Universidad Autónoma Metropolitana, Mexico.

cristobal.aguilar@uadec.edu.mx

A. K. Haghi, PhD
Professor Emeritus of Engineering Sciences, Former Editor-in-Chief, *International Journal of Chemoinformatics and Chemical Engineering* and *Polymers Research Journal*; Member, Canadian Research and Development Center of Sciences and Culture

A. K. Haghi, PhD, is the author and editor of 165 books, as well as 1000 published papers in various journals and conference proceedings. Dr. Haghi has received several grants, consulted for a number of major corporations, and is a frequent speaker to national and international audiences. Since 1983, he served as professor at several universities. He is former Editor-in-Chief of the *International Journal of Chemoinformatics and Chemical Engineering* and *Polymers Research Journal* and is on the editorial boards of many international journals. He is also a member of the Canadian Research and Development Center of Sciences and Cultures (CRDCSC), Montreal, Quebec, Canada.

AKHaghi@gmail.com

Contents

Contributors

Rafail A. Afanas'ev
Pryanishnikov All-Russian Scientific Research Institute of Agrochemistry, d. 31A, Pryanishnikova St., Moscow 127550, Russia

Cristóbal N. Aguilar
Research Group on Bioprocesses and Bioproducts, Food Research Department, School of Chemistry, Universidad Autónoma de Coahuila, 25280 Saltillo, Coahuila, México

Genaro Aguilar-Gutiérrez
Sección de Estudios de Posgrado e Investigación, Escuela Superior de Economía, Instituto Politécnico Nacional, Miguel Hidalgo, Ciudad de México 11340, Mexico

Roberto Arredondo-Valdés
Nanobioscience group, Chemistry School, Autonomous University of Coahuila, Blvd. V. Carranza e Ing. José Cárdenas Valdés, 25280 Saltillo, Coahuila, Mexico

Juan Ascacio-Valdés
Research Group on Bioprocesses and Bioproducts, Food Research Department, School of Chemistry, Universidad Autónoma de Coahuila, 25280 Saltillo, Coahuila, México

J. Basilio Heredia
Centro de Investigación en Alimentación y Desarrollo A.C., Carretera a Eldorado Km. 5.5, Col Campo ElDiez, Culiacán, Sinaloa 80110, México

Diana A. Briceño-Velez
Research Group in Microbiology and Marine Biotechnology - MIBIA. Biology Department, Universidad del Valle, Cali, 25360, Valle del Cauca, Colombia

Luis A. Cabanillas-Bojórquez
Centro de Investigación en Alimentación y Desarrollo A.C., Carretera a Eldorado Km. 5.5, Col Campo ElDiez, Culiacán, Sinaloa 80110, México

Eliseo Cárdenas-Hernández
Research Group on Bioprocesses and Bioproducts, Food Research Department, School of Chemistry, Universidad Autónoma de Coahuila, 25280 Saltillo, Coahuila, México

Mónica Lizeth Chávez-González
Nanobioscience group, Chemistry School, Autonomous University of Coahuila, Blvd. V. Carranza e Ing. José Cárdenas Valdés, 25280 Saltillo, Coahuila, Mexico

Juan C. Contreras-Esquivel
Research Group on Bioprocesses and Bioproducts, Food Research Department, School of Chemistry, Universidad Autónoma de Coahuila, 25280 Saltillo, Coahuila, México

Miriam D. Dávila-Medina
Group of bioprocess and microbial biochemistry, School of Chemistry, Autonomous University of Coahuila, Saltillo 25280, Coahuila, México

Eliseo García-Pérez
Postgrado en Agroecosistemas Tropicales, Colegio de Postgraduados campus Veracruz, Colegio de Postgraduados Campus Veracruz, Veracruz 91700, Mexico

Mayela Govea-Salas
Nanobioscience group, Chemistry School, Autonomous University of Coahuila,
Blvd. V. Carranza e Ing. José Cárdenas Valdés, 25280 Saltillo, Coahuila, Mexico

Arun Kumar Gupta
Department of Food Engineering and Technology, Tezpur University, Tezpur 784028, Assam, India

Yesenia Estrada-Nieto
Nanobioscience group, Chemistry School, Autonomous University of Coahuila,
Blvd. V. Carranza e Ing. José Cárdenas Valdés, 25280 Saltillo, Coahuila, Mexico

Adriana C. Flores-Gallegos
Food Research Department, Universidad Autónoma de Coahuila,
Boulevard Venustiano Carranza S/N CP 25280 Saltillo, Coahuila, México

J. Daniel García-García
Nanobioscience group, Chemistry School, Autonomous University of Coahuila,
Blvd. V. Carranza e Ing. José Cárdenas Valdés, 25280 Saltillo, Coahuila, Mexico

S. M. García-Solares
Centro Mexicano para la Producción más Limpia, Instituto Politécnico Nacional, Av. Acueducto s/n,
Col. La Laguna Ticomán, Ciudad de México 07340, México
Laboratorio Nacional de Desarrollo y Aseguramiento de la Calidad de Biocombustibles (LaNDACBio),
C.P. 07340, Ciudad de México, México

Tabli Ghosh
Department of Chemical Engineering, Indian Institute of Technology Guwahati, Guwahati 781039,
Assam, India

Erick P. Gutiérrez-Grijalva
Cátedras CONACYT-Centro de Investigación en Alimentación y Desarrollo A.C.,
Carretera a Eldorado Km. 5.5, Col Campo ElDiez, Culiacán, Sinaloa 80110, México

C. A. Gutiérrez
Departamento de Ingeniería Química Industrial y de Alimentos (DIQIA),
Universidad Iberoamericana Ciudad de México (UIA). Prolongación Paseo de la Reforma 880,
Santa Fe, Col. Contadero, C.P. 01219, Ciudad de México, México

Ayerim Hernández-Almanza
Group of Bioprocess and Microbial Biochemistry, School of Chemistry,
Autonomous University of Coahuila, Saltillo 25280, Coahuila, México

Catalina J. Hernández-Torres
Bioprocesses & Bioproducts Group, Food Research Department, School of Chemistry,
Autonomous University of Coahuila, Saltillo 25280, Coahuila, México

Anna Ilyina
Nanobioscience group, Chemistry School, Autonomous University of Coahuila,
Blvd. V. Carranza e Ing. José Cárdenas Valdés, 25280 Saltillo, Coahuila, Mexico

Malabika Kalita
Department of Food Engineering and Technology, Tezpur University, Tezpur 784028, Assam, India

Vimal Katiyar
Department of Chemical Engineering, Indian Institute of Technology Guwahati, Guwahati 781039,
Assam, India

Anatolii A. Kovalenko
Pryanishnikov All-Russian Scientific Research Institute of Agrochemistry, d. 31A,
Pryanishnikova St., Moscow 127550, Russia

Liliana Londoño-Hernandez
Group of Bioprocess and Microbial Biochemistry, School of Chemistry,
Autonomous University of Coahuila, Saltillo 25280, Coahuila, México
Research Group in Microbiology and Marine Biotechnology - MIBIA. Biology Department,
Universidad del Valle, Cali, 25360, Valle del Cauca, Colombia

Alejandra Ramírez Martínez
Postgrado en Agroecosistemas Tropicales, Colegio de Postgraduados campus Veracruz,
Colegio de Postgraduados Campus Veracruz, Veracruz 91700, Mexico

José Luis Martínez-Hernández
Nanobioscience group, Chemistry School, Autonomous University of Coahuila,
Blvd. V. Carranza e Ing. José Cárdenas Valdés, 25280 Saltillo, Coahuila, Mexico

Manisha Medhi
Department of Food Engineering and Technology, Tezpur University, Tezpur 784028, Assam, India

Poonam Mishra
Department of Food Engineering and Technology, Tezpur University, Tezpur 784028, Assam, India
E-mail: poonam@tezu.ernet.in

Kona Mondal
Department of Chemical Engineering, Indian Institute of Technology Guwahati, Guwahati 781039,
Assam, India

E. E. Neri-Torres
Departamento de Ingeniería Química Industrial y de Alimentos (DIQIA),
Universidad Iberoamericana Ciudad de México (UIA). Prolongación Paseo de la Reforma 880,
Santa Fe, Col. Contadero, C.P. 01219, Ciudad de México, México

Sandra Palacios-Michelena
Nanobioscience group, Chemistry School, Autonomous University of Coahuila,
Blvd. V. Carranza e Ing. José Cárdenas Valdés, 25280 Saltillo, Coahuila, Mexico

I. R. Quevedo
Departamento de Ingeniería Química Industrial y de Alimentos (DIQIA),
Universidad Iberoamericana Ciudad de México (UIA). Prolongación Paseo de la Reforma 880,
Santa Fe, Col. Contadero, C.P. 01219, Ciudad de México, México

Cristina Ramírez-Toro
Research Group in Microbiology and Marine Biotechnology - MIBIA. Biology Department,
Universidad del Valle, Cali, 25360, Valle del Cauca, Colombia

Rodolfo Ramos-González
Nanobioscience group, Chemistry School, Autonomous University of Coahuila,
Blvd. V. Carranza e Ing. José Cárdenas Valdés, 25280 Saltillo, Coahuila, Mexico

Yadira K. Reyes-Acosta
Bioprocesses & Bioproducts Group, Food Research Department, School of Chemistry,
Autonomous University of Coahuila, Saltillo 25280, Coahuila, México

Ramses M. Reyes-Reyna
Food Research Department, Universidad Autónoma de Coahuila, Boulevard Venustiano Carranza S/N
CP 25280 Saltillo, Coahuila, México
Nanobioscience Research Group, School of Chemistry, Universidad Autónoma de Coahuila,
Boulevard Venustiano Carranza S/N CP 25280 Saltillo, Coahuila, México

Raúl Rodríguez-Hererra
Food Research Department, Universidad Autónoma de Coahuila, Boulevard Venustiano Carranza S/N
CP 25280 Saltillo, Coahuila, México

Aidé Saénz-Galindo
Polymers Department, School of Chemistry, Autonomous University of Coahuila, Saltillo 25280,
Coahuila, México

José Sandoval-Cortés
Analytical Chemistry Department, School of Chemistry, Autonomous University of Coahuila,
Saltillo 25280, Coahuila, México

Elda P. Segura-Ceniceros
Nanobioscience Research Group, School of Chemistry, Universidad Autónoma de Coahuila,
Boulevard Venustiano Carranza S/N CP 25280 Saltillo, Coahuila, México

Anurag Singh
Department of Food Science and Technology, National Institute of Food Technology Entrepreneurship
and Management (NIFTEM), Sonipat, Haryana, India

Dhruv Thakur
Department of Food Science and Technology, National Institute of Food Technology Entrepreneurship
and Management (NIFTEM), Sonipat, Haryana, India

Cristian Torres-León
Research Group on Bioprocesses and Bioproducts, Food Research Department, School of Chemistry,
Universidad Autónoma de Coahuila, 25280 Saltillo, Coahuila, México

Alejandra I. Vargas-Segura
Postgraduate in Advanced Prosthodontics, Dentistry School, Universidad Autónoma de Coahuila,
Boulevard Venustiano Carranza S/N CP 25280 Saltillo, Coahuila, México

Monica Yumnam
Department of Food Engineering and Technology, Tezpur University, Tezpur 784028, Assam, India

Gabriela Vazquez-Olivo
Centro de Investigación en Alimentación y Desarrollo A.C., Carretera a Eldorado Km. 5.5,
Col Campo ElDiez, Culiacán, Sinaloa 80110, México

Tatiana M. Zabugina
Pryanishnikov All-Russian Scientific Research Institute of Agrochemistry, d. 31A, Pryanishnikova St.,
Moscow 127550, Russia

Abbreviations

AD	anaerobic digestion
ADCPT	atmospheric dielectric barrier discharges cold plasma treatment
AP	active packaging
CACM	Mexico City's Wholesale Market
CD	convection drying
CH	cross-linked chitosan
CHN	carbon hydrogen nitrogen
CMC	carboxymethyl cellulose
CNC	cellulose nanocrystals
CP	cold plasma
DBD	dielectric barrier discharge
DH	dielectric heating
DSPME	dispersive solid-phase microextraction
EAE	enzyme-assisted extraction
EOPO	ethylene oxide propylene copolymer of oxide
ET	equivalent Trolox
FAO	Food and Agriculture Organization
FD	freeze-drying
FL	food losses
FLW	food loss and waste
FSC	food supply chain
FTIR	Fourier transform infrared spectroscopy
GAE	gallic acid equivalents
GP	grape pomace
HIU	high-intensity ultrasound
HVED	high voltages electrical discharges
IH	induction heating
IPN	interpenetrating polymer network
LDPE	low-density polyethylene
MAE	microwave-assisted extraction
MAP	modified atmosphere packaging
ML	malolactic
MNP	magnetic nanoparticles

OH	Ohmic heating
PEG	polyethylene glycol
PGG	Penta-galloyl glucose
PHAs	polyhydroxyalconates
PL	photoluminescence
QY	quantum yield
PMA	polymethacrylic acid
PVP	polyvinylpyrrolidone
RNS	reactive nitrogen species
ROS	reactive oxygen species
SCFA	short-chain fatty acid
SEM	scanning electron microscopy
SFT	supercritical fluid technology
SFTE	SFT extraction SWCNTs single-walled carbon nanotubes
SPE	solid-phase extraction
TEM	transmission electron microscopy
UAE	ultrasound-assisted extraction
UV–Vis	ultraviolet–visible
WAE	water bath-assisted extraction
XRD	X-ray diffraction

Preface

This book focuses on the crucial sustainability challenge of reducing food loss at the level of consumer society. Providing an in-depth, research-based overview of the multifaceted problem, it considers environmental, economic, social, and ethical factors. The continuously increasing human population has resulted in a huge demand for processed and packaged foods. As a result of this demand, large amounts of water, air, electricity, and fuel are consumed on a daily basis for food processing, transportation, and preservation purposes. Although not one of the most heavily polluting, the food industry does contribute to the increase in volume of waste produced as well as to the energy expended to do so.

Perspectives included in the book address households, consumers, and organizations, and their role in reducing food loss.

Chapter 1 demonstrates that proper physical and chemical properties, preparation of a function specific hydrogel requires various scientific approaches. Mostly used techniques for producing hydrogel networks are solution and suspension polymerization. They provide molecular structural control, response to stimuli, mechanical strength, biodegradation, and solubility.[1]Earlier, much attention was given to synthetic polymeric materials. But with time, natural materials have grasped the attention of researchers toward them as a result of their amazing properties. Another major factor for their growth is the negative effects of synthetic plastic materials on environment, health, and ecology.

In the food industry, packaging materials have long been used to prevent it from being spoiled. The main reasons for food spoilage are: (a) physical (b) chemical, or (c) biological. This leads to a faster deterioration of food, thereby leading to an increased wastage of food. If food is not properly packed then its chances of spoilage are increased. A good packaging material helps protect from various external factors such as water, oxygen, microorganisms, etc. which may cause harm to food. The use of edible hydrogels as a packaging material has a very wide scope which not only helps in reducing food loss, but also protects the environment as they are biodegradable or can be consumed itself. While edible polymers may have some problems, such as low mechanical strength, flexibility issues, and polymer diffusion; chemical modification or blend forming may overcome these drawbacks and alter their properties in accordance with the application requirements. This provides an advantage over traditional, synthetic polymers which are not edible and may even take up to thousands of years to be degraded, thereby causing a concern for the environment. Different researches are being one on the various natural sources for edible packing which have a huge demand not only from the consumers' side, but also from an environmentalist point of view.

Encapsulation is a widely utilized food preservation technique to entrap various active food ingredients (core materials) within carrier materials to protect it from various external agents. This specific practice is considered as a potential candidate to provide efficient delivery of bioactive compounds into a food system, which further helps in the minimization of food waste. It is noteworthy to mention that there are many advantages of this specific technique to protect and preserve the food products such as it acts as a barrier for the core materials (active agents) against external environment including gaseous agents, light, heat, etc. and the technique also helps to provide tailored-made food property with the aid of encapsulating agents (external phase material). The external phase materials for the encapsulation of a food compound should be food grade, biocompatible, and have the capacity to form a barrier between the external environment and the internal phase. The various available coat materials include carbohydrates (cellulose and derivatives, chitin and chitosan, starch, agar, alginate, carrageenan, gums, and pectin), proteins, lipid, and waxes. Chapter 2 details the various available materials as encapsulating agents (polysaccharides, proteins, and lipids) to encapsulate antioxidants, artificial sweeteners, flavoring agents, and others. Additionally, the various available techniques including spray drying, spray cooling, spray chilling, freeze drying, melt extrusion, melt injection, etc. are widely utilized to develop encapsulates of food products. The chapter also focuses to discuss the details of the experimental procedures and related advantages and shortcomings of the existing techniques of encapsulation.

Chapter 3 deals with sod-podzolic clay loamy soil, acidic, poor available forms of phosphates. The primary condition for increasing its fertility and fertilizer efficiency is the application of phosphorus fertilizers and liming. As the degree of soil culturality is increased due to liming, application of phosphorus fertilizers and manure, the order of nutrient minima changed. In the first and second rotations of the seven-field crop rotation, phosphorus was in the first minimum for all crops, in the third rotation the efficiency of certain types of fertilizers was equalized, and in the fourth rotation nitrogen and potassium came out in the first place, phosphorus took the third place.

Chapter 4 deals with sod-podzolic medium loamy soil of medium degree of cultivation (60–80 mg/kg P_2O_5) under favorable weather conditions mineral and organo-mineral fertilizer systems provided a sufficiently high yield of potatoes, winter wheat, and perennial legumes and grasses. During crop rotation, organo-mineral systems were not inferior in productivity to NPK-equivalent mineral fertilizer systems. In a crop rotation with two fields of perennial legumes and two fields of potatoes, potassium was most effective, followed by phosphorus and nitrogen. On highly cultivated sod-podzolic soil (150–200 mg/kg P_2O_5 and more), the yield of winter wheat, potatoes, barley, hay vetch-oat mixture, and in general the productivity of crop rotation increased under the influence of nitrogen-potassium fertilizer. The effect of phosphorus fertilizer and manure was negligible.

Utilization of waste is becoming a good option from which numerous useful components can be extracted or synthesized. In this book, Chapter 1 of such application will be discussed. *Citrus limon* peels, one of such kind of waste from which a good source of many bioactive and organic compounds can be extracted. Carbon dots are the nanomaterials which size lies below 10 nm, which can be synthesized from many natural and chemical sources. These carbon dots are rich in many properties like fluorescence, photostability, etc which can be utilized in many fields like sensing, bioimaging. So basically, Chapter 5 synthesis of carbon dots form low-cost raw product, methods of synthesis, and utilization of the same in the sensing purposes of ion like Fe^{3+} in water will be discussed. Also, discussion will be extended to prepare some pH-based paper by the utilization of the same for sensing of pH in both acidic and alkaline medium which can be visually distinguished.

The objective of Chapter 6 is to investigate the obtaining of antioxidant bioactive extracts from the Ataulfo mango seed. The influence of two drying methods (convection and lyophilization) and four extraction techniques (water bath, ultrasound, microwave, and enzymatic) were investigated for their total polyphenol content and antioxidant activity (DPPH and ABTS). The HPLC–MS analysis was also developed to identify the compounds. The results showed that lyophilization drying and ultrasound extraction are the most efficient methods with a phenolic content of 29.45 mg GAE/g and a potent antioxidant activity of 97.47% (DPPH) and 289.87 mg TE/g (ABTS) that was higher than that reported in fruits that are a good source of antioxidants. Enzyme-assisted extraction revealed a high polyphenol content and ABTS antioxidant activity, although DPPH antioxidant activity was lower than ultrasound and microwave. This method is proposed as an interesting field of research. HPLC–MS analysis revealed the presence of

16 compounds belonging to the families of lignans, gallotannins, phenolic acids, and flavonols. Our results demonstrate that with the right technology bioactive extracts with a powerful antioxidant activity from by-products of mango processing can be obtained. The extracts have multiple applications in the food and pharmaceutical industry.

The term Food Loss and Waste (FLW) distinguishes between food losses occurring in the supply chain vs food waste; generated in the distribution and final consumption. The composition of the FLW is comprised by carbohydrates, lipids, and proteins, representing an economic alternative to be used as raw material in different bioprocesses that might generate value-added products. Different authors have proposed that the key solution for the food industry resides in finding the appropriate approaches to valorize the FLWs. Obtaining high value-added products from FLWs aligns with the current concept of sustainable development aiming to achieve food security, environmental protection, and energy efficiency. Chapter 7 makes a compilation of main studies to quantify and valorize food loss and waste within a circular bioeconomy.

There is scarce information on the loss of fruits and vegetables registered in the wholesale market of Mexico City, one of the biggest wholesale markets in the world. Considering the above, it is important to generate data on the losses generated in Mexico City wholesale market and to analyze the economic impact of these losses in the region. To achieve this objective in Chapter 8, we visited the fruit and vegetable sector. We obtained the data of the food loss registered by the administrators of the visited stores, which was obtained by the difference in the weight of the truck that transported the product. We also interviewed persons in charge of transport and the administrators/transporters of the visited stores to assess the measures taken during the transportation and selling of the products. The economic impact was quantified by a formulae adapted for the situation. We also carried out an analysis of the transportation in the food loss. Results confirmed that a great quantity of fruit and vegetables are loss and that this loss has a huge impact in the region. To our knowledge, this is the first time such information will be available to the scientific community and the public.

Cereals are the most cultivated crops around the world, with a total production of around 2,723,878,753 tons by 2017. During the processing of cereals, most industries only use grains for human and animal feeding purposes, and the rest of the plant, namely stalks, husk, and leaves, are usually considered as waste. Agro-industrial waste is not usually adequately treated and discarded. Thus, it is an important source of contamination. As an attempt to alleviate food waste contamination, several studies have been conducted to find an economically profitable use of this waste. Most studies have pointed out that cereal by-products are the main source of nutraceutical ingredients such as dietary fiber constituents like lignin, hemicellulose, and cellulose, among others. The components of dietary fiber have attracted much attention because they are the majoritarian compounds found in food waste and due to their reported prebiotic properties. Chapter 9 aims to comprehensively review the latest reports on the bioactive potential of dietary fiber components obtained from cereal by-products.

Chapter 10 describes wine by-products and its generation, importance, and future trends to its use. During wine elaboration, high amounts of by-products are generated; actually, approximately 40% are approached. Although wine by-products do not comprise any danger, its inadequate management can cause adverse effects on the environment, being toxic for cultivars, water, and soils because of the presence of tannins. These by-products are a natural source of diverse phytochemicals with an important biological activity. These compounds are present majorly in skins and seeds of fruits and are highly valorized due to its effect against degenerative diseases as cancer and cardiovascular illness. Wine by-products have been used in animal feed and for recovering alcohol and tartaric acid. Besides, these can be approached due to the viability to generate high-value chemicals such as oils and extracts which can be used in food industry, health, cosmetic, fertilizers, and energy generation.

The use of modified electrodes showed in Chapter 11 leave to see that electrochemical techniques are cheap, easy, fast, and economic ways to develop methods to be used in food analysis in order to achieve good quality and safe food normativity and quality levels to finally arrive to client satisfaction. It pretends to leave clearly the basic concepts around of electrochemical nanostructured sensors, given a few examples of some papers published until 2019, going from the definitions of electrochemical sensors to practical applications of developed electrochemical nanostructured sensors fabricated with different nanomaterials also described. The described nanomaterials are classified in five different kinds, carbon nanotubes, graphene, metal and metal oxide nanoparticles, biomolecules, and molecularly imprinted electrodes. The results described here left no doubts about the convenience when electrochemical techniques are selected, this compared to another typical analytical ways.

Chapter 12 is focused on the aspects of heat transfer applied in food science and technology. Heat transfer is vital on cooking processes. During food preparation, the energy can be transferred by convection, conduction, and radiation. This chapter describes some forms of heat transfer considering the radiation (microwave and ultrasound) and omics technique as the novel methods for matter transformation during the cooking. Heating is used to destroy microorganisms that can transmit disease or alter food quality, as well as to make food more comfortable to digest. Thus, energy transfer helps to obtain products and systems with new properties, simplifies extraction processes, and matter conversion. Moreover, the advances of nanotechnology provide new nanostructured materials that can be applied as catalysts and supports for the immobilization of enzymes and bioactive compounds. The goal of this review is to analyze the different forms of energy transfer and their application in food processing and extraction of active compounds employed to improve food quality.

In Chapter 13, cold Plasma has been defined as an emerging technology that has gained the attention of the food industry due to its great results inactivating microorganisms, like fungi, bacteria, viruses, and spores. Cold Plasma (CP) has been applied to different food like beef, chicken, fruits, vegetables among others, and has been reported its great results to inactivate microorganisms. Even though microorganism deactivation before the treatment has been positive, there has been an interest in applying this technology in food packages to extend the shelf life of the food once it is out in the market and to preserve it fresh. Different researchers have reported that after cold plasma treatment in food packages has been shown that it is possible to preserve food during more time without changing the quality of the food treated. The use of different gases during the cold plasma treatment to carry out different objectives have been reported, some of them are oxygen (O), hydrogen peroxide (H_2O_2), ozone (O_3), nitrogen (N). However, it is important to know that to produce cold plasma, the way the energy is input to the gas is important. These can be carried out by using barrier discharge, corona, jet configurations, operations parameters such as frequency, voltage, power density, among others. In recent years due to the contamination of the environment, the need to replace or reuse different materials like glass, metals, and plastic has grown. Different types of materials such as polypropylene, low-density polyethylene, clamshell container, high barrier cryovac, among others, have been used as food packages modified with cold plasma. The use of different materials used in the application of plasma has gained interest but steel has a lot of unknown questions to be solved.

With the increase in agro-industrial processes has increased the generation of low-value by-products or wastes, which are generally disposed of inappropriately in landfills or other sites, causing serious environmental damage. These agro-industrial by-products are composed mostly of carbohydrates, proteins, minerals, among others, which could be used to obtain value-added molecules. Therefore, the valorization of these materials becomes a priority, in response to current trends of circular economy that indicate that it is necessary to reuse these by-products, in addition to the economic losses generated by wasting such materials. One of the alternative uses would be to obtain compounds with biological activities of interest such as antioxidants, antihypertensives, anticancer, among others, to be introduced in food matrices, which would allow obtaining foods with functional characteristics, extend the useful life of products, improve sensory characteristics, among others. In recent years, research into the processes of obtaining, separating, and purifying high value-added compounds from agro-industrial by-products has gained a special place in the scientific community. However, most extraction processes are based on the use of solvents, which is contrary to current environmental regulations. Therefore, it is necessary to search for and implement alternative technologies that will efficiently obtain these products. The objective of Chapter 14 is to know the advances in the technologies used for the recovery of compounds and their application in the revaluation of agro-industrial by-products, discussing some technological and economic aspects, and the limitations for the use of these technologies to scales superior to those of laboratory.

CHAPTER 1

Application of Hydrogels in Food Packaging to Protect Food Loss

ANURAG SINGH* and DHRUV THAKUR

Department of Food Science and Technology, National Institute of Food Technology Entrepreneurship and Management (NIFTEM), Sonipat, Haryana, India

Corresponding author. E-mail: anurag.niftem@gmail.com

ABSTRACT

Hydrogels are three-dimensional, hydrophilic, polymeric networks capable of absorbing large amounts of water or biological fluids which may be prepared either by natural or synthetic sources. The issue of food loss and waste (FLW) has been receiving increased attention lately. This interest is valid as about one-third of food produced around the world is either wasted or lost as stated by the Food and Agriculture Organization. Packaging plays a vital role in minimizing the wastage of food by increasing its shelf life and protecting it from external factors. One of the important advances in this field is the use of hydrogels as a packaging material. Since synthetic hydrogels are mostly non-biodegradable and are toxic in nature, natural polymers are majorly used for preparing hydrogels as they are biodegradable and can be used as edible packaging also. Majorly used natural polymers for preparing hydrogels are polysaccharides, proteins and lipids. The chapter discusses various aspects related to the use of these hydrogels in food packaging in order to reduce the food loss.

1.1 INTRODUCTION

In 1936, a paper on methacrylic polymers was published by researchers from DuPont. Poly(2-hydroxyethyl methacrylate), for example, pHEMA has been mentioned in this paper and it was described as a difficult, fragile, and vitreous plastic and was obviously not deemed important. Poly-HEMA was largely overlooked until 1960 after that article. In the presence of water or other solvents, Wichterle and Lim discussed polymerization of HEMA and cross-linking agents. They acquired smooth, water-growing, elastic, and clear gel instead of brittle polymers. As we understand them today, this development has resulted in the contemporary field of hydrogels. After that, over the years, the number of formulations for hydrogel has steadily increased.[1] Hydrophilic group macromolecules such as $-OH$, $-COOH$, $-SO_3H$, $-CONH$, and $-CONH_2$ make up hydrogels, which are integrated and incorporated in the plastic backbones. Some physical properties of hydrogels, when fully hydrated, are similar to that of living tissue and natural rubber. Hydrogels are now regarded omnipresent in biomedical, medicinal, and usable food markets across a wide spectrum. Hydrogels are made either from artificial polymers such as pHEMA or natural ones like polysaccharides (cellulose), lipids (waxes) as well as proteins (milk proteins).[2-9]

Hydrogels are three-dimensional, hydrophilic, polymeric networks capable of absorbing large amounts of water or biological fluids.[10] Scientists have described hydrogels in various ways over the years. Mostly used definition states hydrogel is a water-swollen, cross-linked network of polymers, generated with the help of one or more monomers. Due to their outstanding promise in a broad spectrum of applications, hydrogels have earned significant attention over the previous 50 years. Because of their high water content, they are quite comparable to natural tissue in terms of their flexible nature.[11] They have a strong water affinity, but their chemically or physically cross-linked structure prevents them from dissolving.[9] Hydrogels can degrade and eventually break up and dissolve or become stable chemically. These are referred to as physical or reversible gels when the networks are held together by molecular and/or secondary forces, such as ionic, binding, and hydrophobic forces.[12] Hydrogels' competency of imbibing water is because of the hydrophilic groups linked to a polymeric base, whereas because of the cross-links of network chains it is resistant to dissociation.[11] Hydrogels are more similar to living tissue in their physical properties than any other group of synthetic biomaterials. In general, the relatively high amount of water and smooth, rubbery texture give them a close, superficial connection to a living soft tissue.[13]

Hydrogels can be developed in various forms, including plates, nanoparticles, microparticles, films, and coatings. Thus, they find many different applications in various industries.[14] For physical and chemical hydrogels, there are many distinct macromolecular structures. They include cross-linked networks of linear homopolymers, linear copolymers, and block or graft copolymers; multivalent ion polyion, polyion–polyion or H-bonded complexes; hydrophilic networks stabilized by hydrophobic domains; and interpenetrating polymer network (IPN) or physical blends. Physical forms include (1) solid molded form, (2) pressed powder matrices, (3) micropar-ticles, (4) coatings, (5) membranes or sheets (6) encapsulated solids, and (7) liquids.[12]

To demonstrate proper physical and chemical properties, preparation of a function-specific hydrogel requires various scientific approaches. Mostly used techniques for producing hydrogel networks are solution and suspension polymerization. They provide molecular structural control, response to stimuli, mechanical strength, biodegradation, and solubility.[1] Earlier, much attention was given to synthetic polymeric materials. But with time, natural materials have grasped the attention of researchers towards them as a result of their amazing properties. Another major factor for their growth is the negative effects of synthetic plastic materials on environment, health, and ecology.[15]

1.2 BRIEF OVERVIEW OF FOOD WASTAGE IN INDIA

Food wastage as defined by the Food and Agriculture Organization (FAO) is "wholesome edible material intended for human consumption, arising at any point in the food supply chain that is instead discarded, lost or degrade."[16] In recent years, the issue of food loss and waste (FLW) has received increasing attention as a central feature of the challenges and inefficiencies, which characterize the global food system and consequently its social, economic, and environmental implications. According to FAO of the UN, an approximate one-third of food produced around the world is either wasted or lost. This is unfair in an era in which almost 1 billion people go hungry. FLW is a misuse of the labor, water, power, land, and other natural resources that were used for its production.

India is the world's second-largest population. Government announced in 2012 that nearly 22% of the Indian population lived below the poverty line. According to estimates from the FAO in the report "The State of World Food Security and Nutrition, 2017," 190.7 million people in India are undernourished. This accounts for 14.5% of the Indian population, making India the home of the world's largest undernourished population.

However, the UN reports that almost 40% of India's food is wasted or lost which every year costs 1 lakh crore rupees. It is predicted that food production in India will double in the next 10 years; but a cause for concern is postharvest loss of about 35–40% of the total annual output accounting to 58,000 crore.[17]

One of the ways of minimizing the food loss and wastage is by the use of efficient packaging. To avoid handling and in transit losses, fresh-produced items require appropriate packaging. Lack of novel methods of packaging results in a decline in the quality of food, which also contributes to food waste.[16] Hence, an overall improvement in packaging methods is required which will help in controlling the food wastage of the country. Not only does it help in controlling food wastage, but also provides other functions like containment of food and protection from the external environment. These functions help in proper preservation of food and thus improve the shelf life.

We feel proud because we produce one of the largest quantities of foods in the world but we should also be worried to be the first among all in the wastage as well. To meet global food demand for a growing population, food production must not only be significantly increased, but the reasons of food waste should be addressed.[16] Moreover due to an increased concern and awareness towards the environment, we need a sustainable and ecofriendly way of tackling the food loss. This is where biodegradable and edible packaging comes in. Due to the negative effects of traditional, synthetic polymers, there is a need of natural and biodegradable materials which can substitute them for the purpose of food packaging which not only ensures a decrease in food loss but also help in conserving and protecting the environment.

1.3   CLASSIFICATION OF HYDROGELS

The hydrogel products can be categorized according to the following information on different bases as can be seen from Figure 1.1 as well.

1.3.1   BASED ON SOURCE

Hydrogels are of three types based on their source: *natural, synthetic,* or *hybrid hydrogels.*[1]

1.3.2   BASED ON POLYMERIC COMPOSITION

(a) *Homopolymeric*: Hydrogels originating from a single species are known as homopolymeric hydrogels.[15]

(b) *Copolymeric*: These consist of multiple monomers with a minimum of one hydrophilic element, assembled along the polymer network chain in different configurations (arbitrary, block, or back and forth).[11]

(c) *Multipolymeric*: These are also referred to as IPN hydrogel, a category of hydrogels formed by two separate cross-linked polymer components confined in a grid shape. In semi-IPN hydrogels one component is cross-linked and other is noncross-linked polymer.[18]

1.3.3   BASED ON CONFIGURATION

(a) *Crystalline*: It is a well-ordered polymer.

(b) *Semicrystalline:* A composite mix of crystalline as well as amorphous phases.[18]

(c) *Amorphous*: Amorphous polymers are polymers that in X-ray or electron scattering experiments do not show any crystalline structures. They are glassy, brittle, and ductile and have an unordered arrangement.

1.3.4   BASED ON CROSS-LINKING

Based on their chemical and physical behavior of cross-link junctions, hydrogels are of two types.

(a) *Physical*: It is usually achieved through physical mechanisms like hydrophobic interaction, chain aggregation, crystallization, polymer chain complexion, and hydrogen bonding.[1]

(b) *Chemical*: For the preparation of a chemical hydrogel, covalent cross-linking is used either simultaneously or after polymerization.[1]

There are stable interconnections between chemical cross-linked networks, whereas physical networks have temporary interconnections.[18]

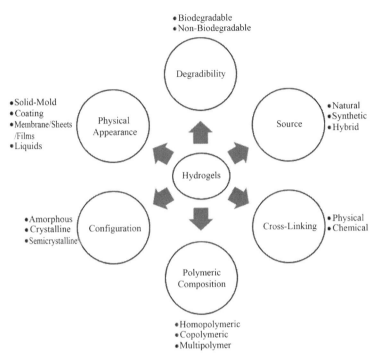

FIGURE 1.1   Classification of hydrogels.

1.3.5   BASED ON PHYSICAL FORM

Hydrogels exist in various physical forms as shown in Figure 1.1. They may appear as solid molds (e.g., contact lenses), powdered matrix (e.g., pills or capsules for oral ingestion), film, coatings (e.g., on pills or capsules), liquid (e.g., gel formers on cooling or heating) or microsphere (e.g., bioadhesive carriers or treatment of wound) depending on polymerization technique.[11,12]

1.3.6   BASED ON DEGRADABILITY

Hydrogels can be either biodegradable or nonbiodegradable. Generally, natural hydrogels are biodegradable, whereas synthetic hydrogels are nonbiodegradable.

1.4 METHODS OF PREPARATION OF HYDROGELS

The basic principle of hydrogel formation is the cross-linking of polymer chain. They are majorly composed of hydrophilic monomers but can also be made of hydrophobic monomers to control the characteristics for desired applications.[11] Hydrogels can be made either from organic or synthetic polymers. Polymers of synthetic nature are stronger chemically, have good mechanical strength, and are hydrophobic in nature.[18] In fact, a hydrogel is a polymer network linked together to produce the flexible structure. Therefore, any technique to build a cross-linked polymer could generate a hydrogel.[11]

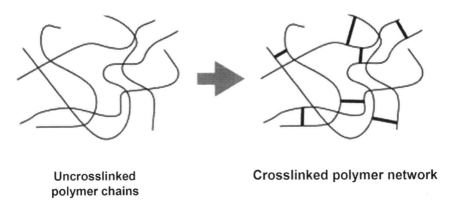

Uncrosslinked polymer chains   **Crosslinked polymer network**

FIGURE 1.2   Cross-linking in polymer.[19]

Natural and synthetic water-soluble linear polymers are interlinked in different ways to form hydrogels: (1) polymer chain interaction by chemical reaction, (2) utilizing ionizing radiation, and (3) physical interaction. The monomer, initiator, and cross-linker are three main components of hydrogel preparation which can be diluted to regulate polymerization heat.[18] New covalent bonds are formed in chemical techniques, whereas there are physical interactions in physically cross-linked hydrogel between polymer chains.[15] Figure 1.3 shows the diagram representing different techniques of hydrogel preparation. Both physical and chemical methods have their own advantages and disadvantages.

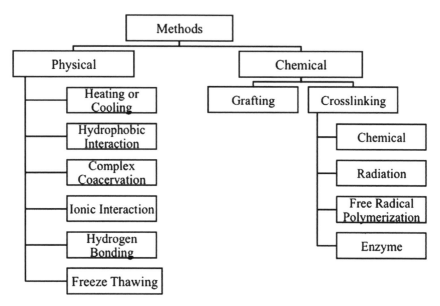

FIGURE 1.3   Methods of preparation of hydrogels.

1.4.1   *PHYSICAL METHODS OF CROSS-LINKING*

Physical cross-links involve various physical interactions among polymer chains other than covalent bonding such as ionic interaction, entangled chains, hydrophobic interaction, hydrogen bonding, and formation of crystallite.[15,20,21] These cross-links are enough to make hydrogels insoluble in aqueous media, though they are not permanent. The major advantage of physical cross-link is it's safety because of the absence of chemical cross-linking agents, thereby preventing the toxic effects from nonreacted chemical cross-linkers.[22]

1.4.1.1   *IONIC INTERACTION*

It involves a physical cross-link with two molecules having opposite electrical charge. Hydrogel cross-linking requires gentle conditions, room temperature, and physiological pH.[15,22] For example, the electrostatic interactions of anionic organophosphorus cross-linkers with cationic lysine residues form elastin-like polypeptides under physiological conditions.[14]

Electrostatic interactions occur between oppositely charged macromolecules which interact with each other to yield polyelectrolyte complexes. The specific advantage of an ionic interaction is its ability to self-heal as a result of which, once the stress is removed, the physical broken network of hydrogels can be reformed. However, due to the cross-linking strategy of the ionic interaction, the mechanical strength of hydrogel is extremely limited.[22] Another disadvantage is the difficulty to control the gelation rate.[15]

1.4.1.2 HYDROPHOBIC INTERACTION

Hydrophobic domain polymers can cross-link through reverse thermal gelation in aqueous environments, also called "sol–gel" chemistry. Hydrophobic interaction occurs at high temperature in an amphiphilic polymer solution. They are soluble at low temperatures in aqueous medium; but its hydrophobic region begins accumulation as temperatures rises to reduce water molecule interaction.[15,22] Hu et al.[22] stated that hydrophobic interactions can be created by two methods: (1) thermal induction based on lower critical solution temperature or upper critical solution temperature and (2) ultrasonic treatment.

1.4.1.3 COMPLEX COACERVATION

It is formed by ionic interactions resulting from the mixing of a polyanion and a polycation.[18] The idea of combining oppositely charged polymers with each other indicates the principle behind this strategy, which renders a pH and concentration-dependent complex,[15] for example, sodium alginate plus polylysine.

1.4.1.4 HYDROGEN BONDING

It is one of the most important noncovalent interactions. These are produced by reducing the pH of the carboxyl-containing polymer solution,[15] for example, complex formation of polyacrylic acid and polymethacrylic acid (PMA) with polyethylene glycol (PEG) from the hydrogen bonds between the carboxylic group of PMA and the oxygen of PEG.[20]

1.4.1.5 FREEZE THAWING

It is also possible to obtain physical cross-linking to form a hydrogel with the help of freeze–thaw cycles. Prior to melting at ambient temperature, the polymer solution is frozen at subzero temperature (-20 to $-80°C$) due to which microcrystals are formed.[15,23]

1.4.2 CHEMICAL METHODS OF CROSS-LINKING

Physically cross-linked hydrogels may be advantageous owing to the absence of chemical agents, but due to uncontrollable factors like gelation time, degradation time, and network pore size, they are somewhat disadvantageous. Their lower mechanical strength is also a major concern.[14,15] On the other hand, chemically cross-linked hydrogels withstand dilution are mechanically stronger and avoid the hydrogel being diffused. These are formed by covalent bonding generally and most of their linkages are permanent and strong unlike physically cross-linked hydrogels. The two important chemical techniques given by Ali and Ahmed[15] as well as Gulrez et al.[23] are grafting and cross-linking.

1.4.2.1 GRAFTING

It includes a monomer's polymerization on the backbone of a preformed polymer. Chemical reagents or treatments with high-energy radiation contribute to the activation of the polymer chain.[23] As suggested by Ahmed[11] this process requires a strong support surface on which free radicals are generated and then polymerization of monomers on it, resulting in covalent bonding of the monomer chain to the support. Grafting can be done in two ways: (1) chemical grafting and (2) radiation grafting. In chemical grafting, molecular backbone activation is done by using chemical initiators such as N-vinyl-2-pyrrolidone, potassium persulfate, benzoyl peroxide, and so on. A 1:2 polymer to monomer ratio showed the maximum grafting rate and water absorption.[15,23] Ionizing rays such as gamma rays and electron beams are used as an initiator for hydrogel formulation of unsaturated compounds. Radicals are formed on the polymer chains due to the polymer solution irradiation. The main benefit of radiation initiation is the development of relatively pure and initiator-free hydrogels.[18] Various advantages of grafting as stated by Ratner and Hoffman[13] are as follows: (a) mechanical strength increase over ungrafted hydrogel, (b) complex surfaces by successive grafting can be obtained using different monomers, (c) using radiation grafting, initiator addition is not necessary, thus a potential source of contamination in final product is eliminated. Some of its disadvantages can be polymer degradation, cross-linking, and unwanted chemical species formation.[13] The grafting efficiency and grafting percentage increased with an increasing monomer concentration and irradiation dose.[23]

1.4.2.2   CROSS-LINKING

It consists of adding into the polymer chain a new functional molecule. For its formation a chemical cross-linker is used (e.g., glyoxal, glutaraldehyde, PEG). However, such cross-linkers escalates hydrogel's toxicity and limits its applicability.[14,15] Thus, using polysaccharides as a base material was reported as a good replacement for chemical cross-linker to make them biocompatible, especially for applications related to biomedicine, agriculture, and food.[15] Radiation and enzyme cross-linking and free radical polymerization are other kinds of chemical cross-linking. Radiation cross-linking is commonly employed as chemical additives are not used due to which biopolymer's biocompatibility is preserved. Modification and sterilization are carried out in one step thereby reducing its cost.[23] Another new way of in-situ hydrogel formation is the enzyme-catalyzed cross-linking reaction, which can serve as a revolutionary alternative to traditional cross-linking methods (e.g., microbial transglutaminases enzyme).[15] Free radical polymerization is achieved through three processes: initiation, propagation, and termination. A free radical active site, which adds monomers in a chain-like fashion, is generated after initiation, for example, poly(N-isopropyl acrylamide) hydrogel.[20]

1.5   HYDROGELS IN FOOD PACKAGING

Packaging is a socioscientific discipline that functions in society to guarantee delivery of goods in the best possible condition.[24] Its major functions are (a) protection, (b) preservation, (c) containment, and (d) convenience. Food storage and preservation is an essential part of food manufacturing. Reduction of risk of contamination or food deterioration and damage, and hygienic delivery of food are important aspects of a good package. A good material provides a good vapor barrier, gas barrier, mechanical properties, aroma retention, flavor retention, enhances microbial stability of food, is environment friendly, and so on. Presently, the main source of material having good properties for packaging are plastics.[15] These pose a major problem of nonbiodegradability and are not environment friendly. Thus, a more ecofriendly way of packaging is required which not only helps in protecting the food but also in protecting the environment.

Synthetic hydrogels are mostly nonbiodegradable, and due to some chemical traces, they may simulate irritation and toxicity.[15] Owing to their biodegradable nature, natural polymers are used to prepare hydrogels. A major advantage of such hydrogels is their edible nature which means that they can be consumed. Edible polymers are the polymeric substance that can be easily ingested in whole or in part by humans or lower animals through the oral cavity and have harmless health effects. These can be applied directly to the surface to maintain the consistency and stability of the material.[25] To be accepted, an edible film must have generally recognized as safe status and used within the limitations set by the US Food and Drug Administration.[26] Environmental friendly packaging solutions are being developed by various groups around the world. Thus, natural polymers are good substitute of traditional synthetic polymers with the advantage of being biodegradable and biocompatible which leads to a decrease in waste and pollution. Though their mechanical properties are not as good as synthetic polymers, but by using a biocross-linker or plasticizer, it can be improved. Hydrogels' characteristics are very much affected by the nature of the polymer. Linear polysaccharides having an inflexible nature with protein form membranes, coatings, and sheets, whereas elastic and globular ones result in film type hydrogels.[15] The main sources for edible polymers according to Theeranun Janjarasskul and John M. Krochta[5] are polysaccharides, proteins, and lipids as shown in Figure 1.4.

Starch can be used to form edible hydrogels which are odorless, transparent, tasteless and above all have low cost, and high environmental appeal, thus do not change the taste, flavor, and visual appearance of food products.[26] Cellulose is also used for production of edible packaging. Edible coating from cellulose ethers, such as methyl cellulose, hydroxypropylmethyl cellulose, and carboxymethyl cellulose (CMC) have been used for packing different foods to provide vapor, oxygen or oil barrier.[15] Because of their relative abundance and strong film-forming performance, protein-based films have been widely used as they have high nutritional value, and provide desirable mechanical properties, gas barriers, and transparency.[27] Total milk proteins or components of milk proteins can form edible films and coatings which act as a protective layer on foods or in between food components.[8] A study done by Sharma and Singh[7] showed that sesame protein can also be used to produce edible films, which are comparable to edible films made by other proteins with thermal and water barrier properties better than other sources such as peanut, soy, and lentil protein. Whey protein-based films have clear appearance, are odorless, and have a good oxygen barrier.[27] Lipids, due to their hydrophobic nature, are used for edible coating to form a water vapor barrier.[5,28] Mostly used substances of this class are waxes.[4,6] Some of the waxes used are beeswax, carnauba wax, rice bran wax, and so on.

Packaging material of edible nature made up of a single source is not as effective as when they are combined. Thus, edible composite packaging materials are created by mixing biocomponents for specific applications with the goal of exploiting complementary functional properties or overcoming their respective flaws.[5] Polysaccharides have good barrier properties but are a poor shield against moisture. While proteins have good mechanical properties and lipids are an excellent repellent of water.[15] To compensate for the weaknesses of one source, a package may be made from more than one source. As stated by Shit and Shah[25] the inclusion in the cellulose ether matrix of hydrophobic compounds such as fatty acids would be a way to improve the moisture barrier, thereby making up for the weakness of poly-saccharides' low resistance to moisture. The hydrophobic nature of zein and its ability to cover surfaces has attracted several industrial applications. By adding pectin to 85% of ethanol which contains zein and calcium chloride, pectin/zein hybrid hydrogels were prepared.[9]

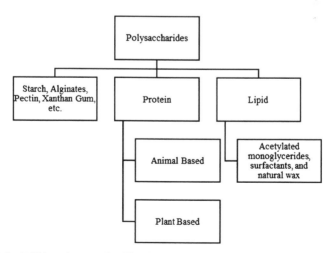

FIGURE 1.4   Edible polymers classification.

Hydrogels can also be used as moisture-absorbent system, with water removal characteristics, which reduce microbial contamination as well as helps in preservation.[29] In a study conducted by Roy et al.,[2] using hydrogels in food packaging systems is an emerging and exciting field where hydrogels are advanced, environment friendly materials capable of maintaining food products' shelf-life. According to Ghosh and Katiyar,[3] hydrogels can be incorporated with active compounds which not only have antimicrobial and antibacterial functions but also do not have any harmful effect on the food components in the package. Moreover, the application to edible films or coatings of food additives such as antimicrobials, antioxidants, flavors, fortified nutrients, and colorants improves its versatility and applicability.[15] In the food and beverage industry, the use of polyvinylpyrrolidone (PVP) and CMC in the production of hydrogel plays an important role. In this regard, hydrogel films based on PVP–CMC have many important properties that are necessary for the application of packaging, including durability, clarity, water-holding capability, and breathability, thus preventing spoilage of food products for a long time.[3] Because of their biopolymeric existence, many hydrogels are biodegraded because they consist of biopolysaccharides and are formed by glycosidic bonds which are degraded by enzymes (chitosan is one such biopolymer).[30] Also, besides such use in packaging, many researchers are also finding potential applications for intelligent as well as active packaging system which interact and may modify the internal environment or the materials inside the package. They may be used as an indicator of temperature, ripening, freshness, and so on[15] which further helps in preventing food wastage as they are a good indicator of the food material inside the package.

1.6 FUTURE SCOPE

Mostly hydrogels are made up of synthetic polymers. Extensive study is required in the field of natural hydrogels as they are renewable, biodegradable, environment friendly and because of raw material availability, it can be manufactured on a large scale. Moreover, some of them are on par with synthetic materials also. Nonetheless, it is important to evaluate their functionality before competing in the market with conventional materials, such as swelling equilibrium, thereby, avoiding the use of toxic reagents, particularly in the cross-linking process.[30] More consideration should be given to the "organic synthesis" used for hydrogel systems which includes using nontoxic solvents and cross-linkers.[15] Antimicrobial food packaging has also gained interest as it helps into inhibit microbial activity, which helps in increasing its shelf life along with food safety. For example, the significant potential use of lipid-based edible films consisting of hydrocolloids (e.g., starch) is as fresh fruit, vegetables, and bread bags.[6] A consumer demand and shift towards the natural and organic products is also a leading trend which could help for a good future of natural hydrogels. Another possibility is the use of nanomaterials for an improvement of hydrogel performance.[15,30]

1.7 CONCLUSION

In the food industry, packaging materials have long been used to prevent it from being spoiled. The main reasons for food spoilage are (a) physical, (b) chemical, or (c) biological. This leads to a faster deterioration of food, thereby leading to an increased wastage of food. If food is not properly packed then its chances of spoilage are increased. A good packaging material helps protect from various external factors such as water, oxygen, microorganisms, and so on, which may cause harm to food. The use of edible hydrogels as a packaging material has a very wide scope which not only helps in reducing food loss but also protects the environment as they are biodegradable or can be consumed itself. Although edible polymers may have some problems, such as low mechanical strength, flexibility issues, and polymer diffusion; chemical modification or blend forming may overcome these drawbacks and alter their properties in accordance with the application requirements.[15] This provides an advantage over traditional, synthetic polymers which are neither edible and may even take up to thousands of years to be degraded, thereby causing a concern for the environment. Different researches are being one on the various natural sources for edible packing which have a huge demand not only from the consumers' side but also from an environmentalist point of view.

KEYWORDS

- **food loss**
- **food wastage**
- **hydrogel**
- **edible packaging**
- **natural polymers**

REFERENCES

1. Ullah, F.; Othman, M. B. H.; Javed, F.; Ahmad, Z.; Akil, H. M. Classification, Processing and Application of Hydrogels: A Review. *Mater. Sci. Eng. C* **2015**, *57*, 414–433.
2. Roy, N.; Saha, N.; Kitano, T.; Saha, P. Biodegradation of PVP–CMC Hydrogel Film: A Useful Food Packaging Material. *Carbohydr. Polym.* **2012**, *89* (2), 346–353.
3. Ghosh, T.; Katiyar, V. Cellulose-Based Hydrogel Films for Food Packaging. *Polym. Polym. Compos.: Ref. Ser.* **2019**, 1061–1084. https://doi.org/10.1007/978-3-319-77830-3_35.
4. Debeaufort, F.; Voilley, A. Edible Films and Coatings for Food Applications. Edible Films and Coatings for Food Applications, 2009; pp. 135–168. https://doi.org/10.1007/978-0-387-92824-1.
5. Janjarasskul, T.; Krochta, J. Edible Packaging Materials. *Ann. Rev. Food Sci. Technol.* **2010**, *1* (1), 415–448.
6. Lee, S. Y.; Wan, V. C. H. Edible Films and Coatings. *Handbook of Food Science, Technology, and Engineering*; CRC Press: Boca Raton, FL, USA, 2005; Vol 4, pp 2406–2417.
7. Sharma, L.; Singh, C. Sesame Protein Based Edible Films: Development and Characterization. *Food Hydrocolloids* **2016**, *61*, 139–147.
8. Shendurse, A. Milk Protein Based Edible Films and Coatings—Preparation, Properties, and Food Applications. *J. Nutr. Health Food Eng.* **2018**, *8* (2), 219–226.
9. Liu, L. S.; Kost, J.; Yan, F.; Spiro, R. C. Hydrogels from Biopolymer Hybrid for Biomedical, Food, and Functional Food Applications. *Polymers* **2012**, *4* (2), 997–1011.
10. Peppas, N. A.; Bures, P.; Leobandung, W.; Ichikawa, H. Hydrogels in Pharmaceutical Formulations. *Eur. J. Pharm. Biopharm.* **2000**, *50* (1), 27–46.
11. Ahmed, E. M. Hydrogel: Preparation, Characterization, and Applications: A Review. *J. Adv. Res.* **2015**, *6* (2), 105–121.
12. Hoffman, A. S. Hydrogels for Biomedical Applications. Adv. Drug Deliv. Rev. **2002**, *54*, 3–12.
13. Ratner, B. D.; Hoffman, A. S. Synthetic Hydrogels for Biomedical Applications. *Hydrogels for Medical and Related Applications*; American Chemical Society: USA, 1976.
14. Hoare, T. R.; Kohane, D. S. Hydrogels in Drug Delivery: Progress and Challenges. *Polymer* **2008**, *49* (8), 1993–2007.
15. Ali, A.; Ahmed, S. Recent Advances in Edible Polymer Based Hydrogels as a Sustainable Alternative to Conventional Polymers. *J. Agric. Food Chem.* **2018**, *66* (27), 6940–6967.
16. Balaji, M.; Arshinder, K. Resources, Conservation and Recycling Modeling the Causes of Food Wastage in Indian Perishable Food Supply Chain. *Resour. Conserv. Recycl.* **2016**, *114*, 153–167.
17. Negi, S.; Anand, N. Post-Harvest Losses and Wastage in Indian Fresh Agro Supply Chain Industry : A Challenge. *IUP J. Supply Chain Manage.* **2017**, *14*, 07–23.
18. Garg, S.; Garg, A.; Vishwavidyalaya, R. D. Hydrogel : Classification, Properties, Preparation, and Technical Features. *Asian J. Biomater. Res.* **2016**, *2* (6), 163–170.
19. De Azeredo, H. M. C.; Rosa, M. F.; De Sá, M.; Souza Filho, M.; Waldron, K. The Use of Biomass for Packaging Films and Coatings. *Advances in Biorefineries: Biomass and Waste Supply Chain Exploitation*, 2014; pp. 819–874.
20. Maitra, J.; Shukla, V. K. Cross-linking in Hydrogels—A Review. *Am. J. Polym. Sci.* **2014**, *4* (2), 25–31.

21. Oyama, T. Encyclopedia of Polymeric Nanomaterials. *Encycl. Polym. Nanomater.* **2014,** *1,* 1–11.

22. Hu, W.; Wang, Z.; Xiao, Y.; Zhang, S.; Wang, J. Advances in Crosslinking Strategies of Biomedical Hydrogels. *Biomaterials Science*; Royal Society of Chemistry: United Kingdom, 2019; pp. 843–855.

23. Gulrez, S. K. H.; Al-assaf, S.; Phillips, G. O. Hydrogels: Methods of Preparation, Characterisation and Applications. *Progress in Molecular and Environmental Bioengineering—From Analysis and Modeling to Technology Applications*; Carpi, A., Ed.; IntechOpen, London, UK, 2011.

24. Robertson, G. L. *Food Packaging Principles and Practice*, 3rd ed.; CRC Press: Boca Raton, Florida, USA, 2013; pp 1–9.

25. Shit, S. C.; Shah, P. M. Edible Polymers: Challenges and Opportunities. *J. Polym.* **2014,** 1–13.

26. Suput, D.; Lazic, V.; Popovic, S.; Hromis, N. Edible Films and Coatings: Sources, Properties, and Application. *Food Feed Res.* **2015,** *42* (1), 11–22.

27. Said, N. S.; Sarbon, N. M. *Protein-Based Active Film as Antimicrobial Food Packaging: A Review*; Intech, 2016.

28. Kahve, H. I.; Ardic, M. Lipid-Based Edible Films. *J. Sci. Eng. Res.* **2017,** *4* (9), 86–92.

29. Otoni, C. G.; Espitia, P. J. P.; Avena-bustillos, R. J.; Mchugh, T. H. Trends in Antimicrobial Food Packaging Systems : Emitting Sachets and Absorbent Pads. *Food Res. Int.* **2016,** *83*, 60–73.

30. Batista, R. A.; Espitia, P. J. P.; Quintans, J. de S. S.; Freitas, M. M.; Cerqueira, M. Â.; Teixeira, J. A.; Cardoso, J. C. Hydrogel as an Alternative Structure for Food Packaging Systems. *Carbohydr. Polym.* **2019,** *205* (September 2018), 106–116.

CHAPTER 2

Encapsulation: A Customized Practice for Minimization of Food Waste

TABLI GHOSH, KONA MONDAL, and VIMAL KATIYAR*

Department of Chemical Engineering, Indian Institute of Technology Guwahati, Guwahati 781039, Assam, India

Corresponding author. E-mail: vkatiyar@iitg.ac.in

ABSTRACT

Encapsulation is a widely utilized food-preservation technique to entrap various active food ingredients (core materials) within carrier materials to protect it from various external agents. This specific practice is considered as a potential candidate to provide efficient delivery of bioactive compounds into a food system, which further helps in the minimization of food waste. It is noteworthy to mention that there are many advantages of this specific technique to protect and preserve the food products such as it acts as a barrier for the core materials (active agents) against external environment including gaseous agents, light, heat, and so on and the technique also helps to provide tailored-made food property with the aid of encapsulating agents (external phase material). The external phase materials for the encapsulation of a food compound should be food grade, biocompatible, and have the capacity to form a barrier between the external environment and the internal phase. The various available coat materials include carbohydrates (cellulose and derivatives, chitin and chitosan, starch, agar, alginate, carrageenan, gums, and pectin), proteins, lipid, and waxes. The present chapter details the various available materials as encapsulating agents (polysaccharides, proteins, and lipids) to encapsulate antioxidants, artificial sweeteners, flavoring agents, and others. Additionally, the various available techniques including spray drying, spray cooling, spray chilling, freeze drying, melt extrusion, melt injection, and others are widely utilized to develop encapsulates of food products. The chapter also focuses to discuss the details of the experimental procedures and related advantages and shortcomings of the existing techniques of encapsulation.

2.1   INTRODUCTION

Encapsulation is considered as a customized practice for developing functional food products, where active substances are generally entrapped within a carrier material. The entrapped substances within the carrier materials are known as "internal phase" or "core material" and on the other hand, the carrier material or encapsulating agents, which are used to develop the outer layer, are known as "external phase" or "matrix materials." The encapsulation is a core–shell system, where the shell materials are carbohydrates, proteins, polymer blends, liposomes, and so on. The core materials are dyes, flavors, bioactive compounds, fats and oils, aroma compounds, oleoresin, vitamins, minerals, colorants, enzymes, and so on. The developed particles as encapsulates have a diameter of around a few nanometer to a few millimeter range range. However, the encapsulation process can be categorized as "microencapsulation" or "nanoencapsulation" where the core substances in the form of micro or nanoparticle are encapsulated within a shell material to form microencapsulates or nanoencapsulates of active compounds. According to the size of encapsulates, it can be classified as microcapsules (1–1000 μm), submicron capsules (several 100 nm to less than 1 μm) or nanocapsules (1 to several 100 nm), which can be produced *via* "microencapsulation" or "nanoencapsulation" technique.[1] The principle focuses of encapsulation techniques are to stabilize an active ingredient with a control release mechanism, and conversion of solid/liquid/gaseous components into a solid, which can be handled easily. Further, the processing steps of encapsulation process involve the selection of some parameters such as selection of wall materials, selection of core materials, preparation of solution or emulsion (optimization of the proportion of wall and core material), selection of encapsulation technique as shown in Figure 2.1. The encapsulating agents that are used to prepare encapsulates are generally food-grade materials and have the potential to act as a barrier between active agents and the surrounding. It is noteworthy to mention that the encapsulation of food components provides various advantages, such as reduced food waste, protect active compounds, provide control release of encapsulated materials, develop functional food products, provide nutritional benefits, and health benefits, and so on.

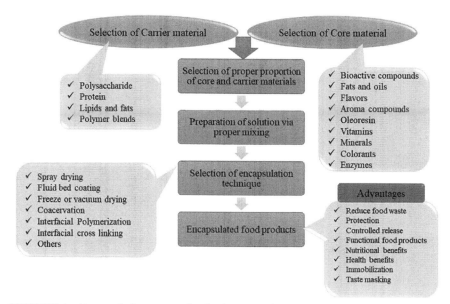

FIGURE 2.1   Encapsulation process for food preservation.

Additionally, the structures of encapsulates in terms of size, shape, internal designs are essential factors to obtain effective encapsulate materials for targeted applications. The encapsulated materials are generally of two types such as reservoir type and matrix type, where the reservoir type may have a single coat or multilayer coat around the active agent and this single-coat material is generally known as single-core, mono-core, and core–shell type. On the other hand, in the matrix type encapsulates, the core materials remain as a disperse phase over the encapsulating agents, and the active agents can be in the form of small droplets or can be distributed over the carrier material. In the matrix type of encapsulation, the active agents are obtained at the surface or they can also have an additional coating material as shown in Figure 2.2. Further, the encapsulated materials can also be categorized in terms of shapes such as simple, multiwall, multicore, matrix, irregular, and so on as shown in Figure 2.3. Interestingly, the core materials sometimes may form network-like structure with the external phase materials.

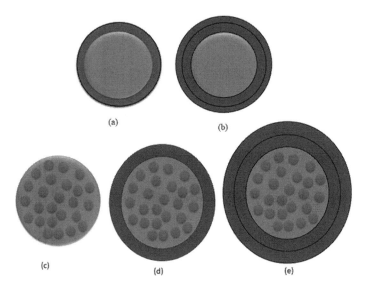

FIGURE 2.2   Various forms of encapsulations: (a) reservoir type, (b) multilayer reservoir type, (c) matrix type, (d) coated matrix type, and (e) multilayered matrix type.

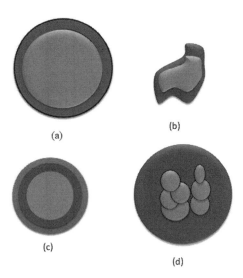

FIGURE 2.3   (a) Simple, (b) irregular, (c) multiwall, and (d) multicore type of encapsulation.

Based on the above discussion, the chapter focuses to discuss the significance of encapsulation technique in regards to reduce food loss. Moreover, the available materials used for the encapsulation techniques will also be detailed in this chapter. The encapsulation of various food ingredients is obtained *via* various techniques such as spray drying, freeze drying, spray cooling, ionic gelation, solvent evaporation extraction, coacervation, and others. The details about the various available encapsulation techniques will also be discussed in this chapter.

2.2 SIGNIFICANCE OF ENCAPSULATION FOR MINIMIZATION OF FOOD WASTE

As discussed in the previous section, the encapsulation approach has several benefits for the minimization of food waste by developing a barrier between the core food material and the surrounding, which helps in removing the bad tasting and smelling, and further increase the bioavailability of food ingredients. Additionally, the encapsulation of food products provides many other advantages as shown in Figure 2.4, which includes protection toward sensitive food products such as active compounds, oils, vitamins, minerals, and others. The encapsulated food products can be incorporated into solid or liquid food products to develop functional or fortified food products with nutrient-rich foods. The process also helps in providing improved stability during food processing, where the active compounds undergo reduced degradation. The barrier also helps in reduced impact of environmental agents such as gaseous environment, light, water, and others. The encapsulation technique also modifies the physical properties of original active materials; provides controlled release, immobilization, easy handling, storage, and acceptability; improves safety, textural effects, tailored aroma properties, such as particle size, structure, controlled release; and so on. Moreover, the encapsulation techniques are used to improve the flow properties, enhanced stability, reduced volatility of materials, gastric irritation, and so on. It is noteworthy to mention that encapsulation-based approaches are widely used in functional food ingredients as there is an increased demand for functional food products. The use of encapsulation has many advantages such as increased nutritional and health benefits, unaltered taste, increased shelf life, and further, encapsulated materials don't interfere with other ingredients. The developed encapsulated materials can also be added at any time of processing of food products for improved functionality.

The global food encapsulation market by technology includes microencapsulation, nanoencapsulation, macroencapsulation, and hybrid technology. However, the various functionality of microencapsulation includes masking undesirable flavors, improve flow properties, increased shelf life, controlled release profile, protect bioactive compounds, product enrichment with specific nutrients, whereas the functionality of nanoencapsulation includes higher surface area, improved delivery of bioactive compounds, improved bioavailability, increased shelf life, improved intracellular uptake, and so on.[1] In this regard, the application area of encapsulation in food products includes oxidation and flavor stabilization, taste masking, color masking, and so on. The encapsulation technique is generally used in pharmaceutical, cosmetics, foods and printing industries, chemical industries, and so on as shown in Fig. 2.4. The encapsulated materials are widely used in pharmaceuticals industry for obtaining controlled and targeted drug delivery of components. Further, the entrapped core materials are generally released by various approaches such as pH, diffusion, and degradation, which further depends on the stability of the capsule. The triggered release of core or active materials can be obtained *via* optimized pH, temperature, and pressure. Further, targeted release and sustained release can be obtained *via* site recognition, and *via* shell decomposition, respectively. In this way, tailor-made properties of encapsulates of active compounds can be attained *via* tuning the processing and surrounding conditions.

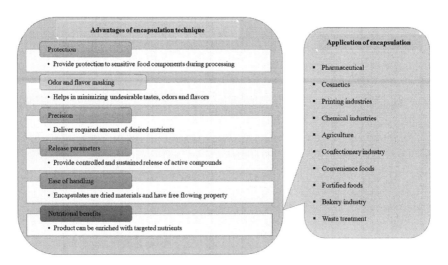

FIGURE 2.4   Advantages and application of encapsulation techniques.

2.3 FOOD-GRADE CARRIER MATERIALS AS ENCAPSULATING AGENTS

The carrier materials for encapsulation are considered as one of the important factors in obtaining the efficient encapsulates of active compounds. There are several materials used for encapsulation technique and possess many properties to be used as an effective material for the active agents. In this regard, the carrier materials should be food-grade materials and have the potential to act as a barrier between active agents and the surrounding. As described in Figure 2.5, the generally used encapsulating agents are biopolymers which are biodegradable, biocompatible, biofunctional, bioavailable, high solubility in water, good film-forming property, good emulsifying property, low cost, provide improved stability, protect the active materials, and so on. The selection of encapsulating agents is generally depending on various properties, such as the nature of core materials, product requirement, encapsulation process, should provide maximum protection of the active ingredients, and so on. For the preparation of encapsulated materials, carbohydrates, proteins, and lipids are used. The physicochemical and rheological behavior of carrier materials and core materials needs to be understood properly before selecting the carrier materials. The available carrier materials include classes and subclasses of polysaccharide, protein, fat, and lipid-based material,[1,2] which will be discussed in the below section. The various food ingredients that are encapsulated include (1) flavoring agents; (2) acids, alkalies, buffers; (3) lipids; (4) redox agents such as bleaching and maturing; (5) enzymes and microorganisms; (6) artificial sweeteners; (7) leavening agents; (8) antioxidants; (9) preservatives; (10) colorants; (11) cross-linking and setting agents; (12) agents with undesirable flavors and odors; and (13) essential oils, amino acids, vitamins, and minerals, and so on.[3,4] Further, anthocyanin, polyphenolic compounds, essential oils, oleoresins are the focused active compounds which are widely encapsulated to protect the beneficial agents. Anthocyanin can be easily availed from red onions, pomegranates, grapes, tomatoes, berries, chokeberry, tart cherries, and others. Anthocyanin has many beneficial properties for humans, animals, and plants and can be used as a dietary supplement as anthocyanin-rich food products. This specific pigment generally belongs to a class of polyphenols, which has the characteristic features of a broad colored scheme from bright red to dark blue and purple color and are water-soluble pigments.[5] Anthocyanin is very sensitive toward environmental conditions such as gaseous conditions, temperature, pH, light, and so on. Thus, this pigment needs to be protected using encapsulation techniques for efficient delivery to other food products. On the other hand, essential oils can be encapsulated to provide optimized functionality, where the activity of essential oils can be stored for a prolonged time.[6] The essential oils can be extracted from fruits, spices, plants, and others. The essential oils have the properties of antimicrobial, antibacterial, analgesic, disinfectant, and others. The various available essential oils include basil-based, bergamot-based, black pepper-based, camomile-based, cedarwood-based, cinnamon-based essential oils, and so on. Polyphenolic compounds are highly available from plants and have medicinal and industrial values. Further, food waste is also a rich source of bioactive compounds, which can also be utilized to develop functional food products. The various waste materials obtained from agricultural, industrial, household, and other food sectors can be utilized to obtain bioactive compounds. The generally available food wastes include cereals, fruit, and vegetable wastes, which are utilized widely to obtain health beneficial nutraceuticals, bioactive compounds, and helps in reducing food waste. However, sustainable nanostructured materials are also used for encapsulation of active compounds. Interestingly, various available organic and inorganic sustainable nanostructured materials such as nanocellulose, nanochitosan (can be available from waste materials), and others are extensively applied in the area of food packaging applications for improved properties and functionality.[7] In this regard, the various available materials used for encapsulation of discussed materials will be made in this section.

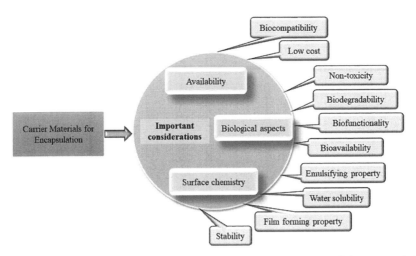

FIGURE 2.5 An important consideration for carrier materials used in encapsulation techniques.

2.3.1   CARBOHYDRATES AS A CLASS OF ENCAPSULATING AGENT

Polysaccharides are defined as a class of carbohydrates consisting different kinds of monosaccharide units linked by glycosidic bonds. The carbohydrates can be obtained as monosaccharides, disaccharides, oligosaccharides, and polysaccharides. Carbohydrates can be obtained from various plant-based sources (cellulose, pectin, starch, gum arabic, maltodextrin, dextrin, and cyclodextrin), animal-based sources (chitosan), microbial sources (xanthan, dextran, pullulan, and cyclodextrin), and algae sources (alginate, carrageenan). Carbohydrates provide the benefits of low cost, widely used in food products, good solubility, low viscosity, bind flavors, and so on. The significance of using carbohydrate-based materials for encapsulation includes high stability, wide availability, low cost, whereas the materials have the disadvantage such as number of reactive functional groups, and high energy is required to obtain nanoscale materials.

2.3.1.1   CELLULOSE AND ITS DERIVATIVES

Cellulose is a group of polysaccharide units having β-linked D-glucose units as a repeating monomer and is the most available form of biopolymer.[8] Cellulose is a renewable, widely available, nontoxic, and most investigated biopolymers, which is used to develop encapsulation-based materials. Cellulose has a tunable functional property due to its surface chemistry, which provides thermal gelation, surface modification, solubility to deliver barrier property, controlled delivery of active compounds, and stabilization in formulations, and so on. Cellulose nanocrystals (CNC) are also extensively utilized for microencapsulation purpose, where CNC cross-linked chitosan (CH) have a better encapsulation efficiency in comparison to CNC tripolyphosphate complex.[5] In this regard, in CH–CNC complex, CNC works as a filler material in CH-based matrix to develop a stable microcapsule. Further, cellulose ethers and esters are extensively utilized for dietetic foods, fried foods, bakery foods, extruded foods, frozen desserts, heated convenience foods, and others.[9] The cellulose ethers after thermal gelation having characteristics features of hydrophilicity, water-containing gels, which can be used as a food coating and batter, and have the ability to limit oil penetration property during frying food products. Further, hydroxypropyl cellulose as temperature responsive gels is used for encapsulation of food products and has the capacity of retaining flavor at cooking temperature.[10] The modified forms of cellulose such as methylocellulose and hydroxypropyl methylocellulose has a good emulsifying property, which can be used as a coating material for developing spray-dried fish oil for improved stability of fish oil.[11] The acetophtalate cellulose is another derivative of cellulose, and widely used in industry; however, acetophtalate cellulose possesses some disadvantages such as synthetic, slow process of drug absorption, and it dissolves only above pH 6.[12] The other form of cellulose such as ethyl cellulose is derived from glucose and is insoluble in water, whereas having some disadvantages such as insolubility in gastrointestinal system, and low mechanical resistance, and others.[12]

2.3.1.2   STARCH AND ITS DERIVATIVES

Starch is one of the potential biopolymers having the advantages of low cost, nontoxic, abundancy, renewability and is mostly obtained from green plants.[13,14] Starch is a kind of polysaccharide which is consisting of amylose and amylopectin biopolymers associated with α-glycosidic bonds.[1,15,16] The amylose is a linear biopolymer consisting of $\alpha(1\rightarrow4)$-linked glucose units and amylopectin is a branched biopolymer consisting of $\alpha(1\rightarrow4)$-linked glucose units (as linear chains) linked to $\alpha(1\rightarrow6)$-linked glucose units (as branched chain).[13] Starch from different sources has different molecular weights of the two components of amylose and amylopectin, which provide different processability.[13] The starch modification is obtained *via* tailored chemical property through chemical, enzymatic, and physical methods. The starch modification *via* chemical process includes degradation (acid hydrolysis, oxidation, dextrinization, enzymatic treatment), substitution (oxidation, esterification, etherification), and cross-linking. In the chemical modification of starch, there involved modification of covalent linkages, cross-linking of glucan chains and addition of charged group to the chains. The physical modification includes pregelatinization such as roll-drying, spray, thermal treatment, where the factors such as moisture, heat, ultrahigh pressure are involved. The various modified starch components include dextrins, maltodextrins, cyclodextrin, resistant starch, and so on. The dextrins are hydrolyzed starch, which is a group of low molecular weight carbohydrates consisting of $\alpha(1\rightarrow4)$-linked and $\alpha(1\rightarrow6)$-linked D-glucose units. The dextrin obtained by applying dry heat is known as pyrodextrins, which are available as white dextrin, yellow dextrin, and brown dextrin. The brown and yellow dextrins are used as carriers for food products due to their high water soluble nature, whereas yellow dextrin can also be utilized as an encapsulating agent for water-insoluble flavorings. Maltodextrins are a class of dextrin and a hydrolyzed starch, having a dextrin equivalent of less than 20. The maltodextrin has many beneficial properties such as high water solubility, provides low viscosity in solution, reduced agglomeration problems. In the development of encapsulates, maltodextrin is mixed with other agents such as pectins, gums, alginates, and protein-based materials for improved efficiency. Additionally, cyclodextrins are a kind of modified starch developed by enzymatic treatment. The available forms of cyclodextrin include α, β, and γ forms of cyclodextrins, where β-cyclodextrin is widely utilized for encapsulation purpose for its low cost. Additionally, other starch forms such as nanoparticles, iron incorporated starch are widely utilized for encapsulation purposes.

2.3.1.3   CHITOSAN

Chitosan is a deacetylated product of chitin biopolymer, which is consisting of β-(1→4)-linked D-glucosamine and N-acetyl-D-glucose-amine.[1,12,15,16] The chitosan can be obtained from shrimp, lobster, crab, muga silkworm, prawns and has the properties of biocompatibility, nontoxicity, antimicrobial, antibacterial, antidiabetic, anticholesterolemic, hypocholesterolemic, antioxidant properties, and so on.[15,17] The advantages of chitosan to be used as an encapsulating agent include naturally available, nontoxic, degradable, biocompatible, and provide a gradual release of various drugs. Further, chitosan has an ability to adhere to gastric mucose, and provide no allergic reactions, which make it an essential agent for several versatile applications. Chitosan can be used as a matrix and other cross-linking agents can be used with chitosan for having the amino group attached with it.[5] The various cross-linking agents used for developing chitosan-based encapsulations include citrate, sulfate, phosphate, tripolyphosphate, and so on. Chitosan has positive charges, so it can react with polyanionic compounds as cross-linking agents to develop complex which can carry active substances easily. The acetylation of chitosan makes them positively charged and the anionic cross-linking agents can create active compounds loaded microcapsules. The chitosan-based microcapsules are widely utilized with other cross-linking agents such as CNC and sodium tripolyphosphate, and others.[5] Further, chitosan, sodium hydroxide, and surfactants are used to microencapsulate limonene essential oils using coacervation technique, where the controlled release of active agents can be obtained *via* tunable sodium hydroxide concentration.[6]

2.3.1.4   OTHER COMPONENTS

The other components to be used as encapsulating agents include gums, pectin, alginate, and so on. The available gums for encapsulation include gum arabic, gum acacia, alginates, carrageenan, and others. Gums have a property to provide viscosity and gelling ability to food products and are generally obtained from seaweeds, seeds, microbes, and others. The gums are water soluble and highly hydrated. Gum arabic is a kind of polysaccharide consisting of β-1,3-glycoside-connected galactose unit and provides characteristic features of antimicrobial and antioxidant property.[17,18] Gum arabic has the advantages of adequate solubility, low viscosity, smooth taste, good emulsifying properties, and various other functions. On the other hand, gum arabic has the disadvantages of high cost and low availability, whereas alginate is a kind of natural product and easily working materials. The disadvantages of alginate are high cost at the level of industrial scale and have a porous and permeable membrane property. In addition to these materials, there are many available materials, which are used to develop encapsulates of active compounds.

2.3.2   PROTEINS AS A CLASS OF ENCAPSULATING AGENT

Proteins are considered as a potential candidate to be used as encapsulating agents. Both animal- and vegetable-derived proteins are used for encapsulation of active substances. The advantages of using protein-based carriers include nutritional benefits, good emulsifiers, gelation properties, and others. The various protein sources include plant origin such as zein, soy-protein isolate, and animal-based proteins such as whey-protein isolate, whey-protein concentrate, egg albumin, collagen, gelatin, silk fibroin, and so on. The plant-derived proteins have achieved a greater interest in comparison to animal-derived proteins due to their good hydrophobic nature, less expensive, less allergic nature, and other advantageous properties. The proteins are considered as very effective encapsulating agents for various active agents such as bioactive compounds, fatty acids, flavors, fats, and oil, and more. Gelatin has the advantages of natural product, low price, nontoxic, and biodegradable, whereas it also has a high solubility in an aqueous system. The use of bioactive protein hydrolysates and peptides whose wide involvement in functional food may have some limitations such as compatibility. Casein is a milk protein, which has the characteristics features such as high nutritional value, low viscosity in solution, good emulsifying property, which make them an excellent carrier material for bioactive agents. The commercially available casein protein includes sodium caseinates and calcium caseinate, which can be obtained by solubilizing native casein. On the other hand, soy proteins are also utilized for encapsulating agents for their emulsification property, gel formability, capability of water binding, nutrient protecting nature, and others. The protein materials in combination with other polysaccharide units are used to obtain the improved property of encapsulated materials.

2.3.3 LIPIDS AS A CLASS OF ENCAPSULATING AGENT

The use of lipid-based carriers has many advantages such as nanocapsules can be achieved with low energy and lipid-based carriers are also suitable for peptides with different properties. However, the use of lipid-based carriers has many disadvantages such as lipid oxidation, thermal instability above-phase transition temperature of liposomes.[19] The various lipid components, which are used as encapsulating agents include hydrogenated fat, waxes, and paraffin, acetoglyserides, shellac resins, glycolipids, fractionated fats, mono- and di-glycerides, natural fats and oils, phospholipids, waxes—beeswax, carnauba wax, emulsifiers, and so on. The lipid-based encapsulating agents include fats and oils containing polar and nonpolar lipids. The lipid-based materials as a carrier material have many advantages such as emulsification property, film-forming property, less toxic, which helps in efficient encapsulation of bioactive compounds. The lipid-based encapsulation has many significant properties such as provide improved encapsulation efficiency, better stability, improved bioavailability, and reduced toxicity. In this regard, some of the various classes of materials for encapsulation purposes are discussed in Table 2.1.

TABLE 2.1   Several Classes of Encapsulating Agents for Active Food Components.

Sl. No.	Encapsulating agents	Internal phase	Property	Reference
1.	Methylocellulose Hydroxypropyl methylocellulose	Fish oil	Improves stability Improves concentration of fish oil in powder	[11]
2.	Chitosan Sodium hydroxide Surfactants	Limonene essential oil	Control release of limonene encapsulated with sodium hydroxide concentration	[6]
3.	Cellulose nanocrystal Chitosan Sodium tripolyphosphate	Fruit anthocyanin	Cellulose nanocrystal CH has better encapsulation efficiency than cellulose nanocrystal cross linked tripolyphosphate	[5]
4.	Temperature responsive hydrypropyl cellulose	Flavor material	Increased release time compared to gelatin capsules	[10]
5.	Maltodextrin	Black carrot anthocyanin	Increased half-life of black carrot anthocyanin (developed by spray-drying process) at 4°C in comparison to 25°C storage temperature	[20]
6.	Soybean protein isolate/pectin	Casein hydrolysate	Provide controlled release of casein hydrolysate Reduced bitter taste Better hydrolysate encapsulation Lower hygroscopicity	[21]
7.	Soybean protein isolate	Casein hydrolysate	Provide a functional food product Attenuate the bitter taste of casein hydrolysate	[22]
8.	Gum arabic Maltodextrin Modified starch	Cumin oleoresin	Gum arabic acts as a better carrier material than other materials	[23]
9.	Gum arabic Modified starch	Flavoring agent: cardamom oleoresin	Gum arabic acts as better carrier materials in comparison to modified starch and maltodextrin	[24]

TABLE 2.1 *(Continued)*

Sl. No.	Encapsulating agents	Internal phase	Property	Reference
10.	Gum arabic Starch Maltodextrin Inulin	Rosemary essential oil	Encapsulated rosemary oil maintains all the main constituents Maltodextrin and modified starch in the ratio of 1:1 found to be a suitable carrier material	[25]
11.	B-cyclodextrin Modified starch	Caraway extract	Protect volatile substance	[26]

2.4 AVAILABLE TECHNIQUES FOR ENCAPSULATION

The several techniques that are available for encapsulation of active molecules in food and pharmaceutical industries are represented in Figure 2.6. In recent days, these techniques have become very much popular and effective for providing beneficial effects to the environment and human beings. Based on availability, advantages, and existing shortcomings of available techniques various types of compounds are being able to encapsulate.

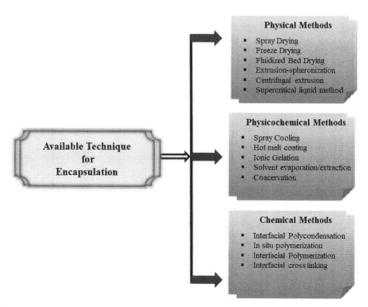

FIGURE 2.6 Available techniques for encapsulation of active compounds.

2.4.1   PHYSICAL METHODS

2.4.1.1   SPRAY DRYING

Spray-drying technique is one of the most available and well-known techniques for food encapsulation. This technique is very much useful in the industrial sector to obtain stable powders from liquids. It is a flexible, continuous, and economically feasible technique, which produces particles of good quality with a size of fewer than 40 μm.[27] In general, most of the microencapsulates are developed by spray-drying technique.[28] This process commonly includes some of the stepssincluding dissolving the core materials and carrier or wall materials in a selected solvent which helps to form an emulsion or dispersed solution followed by atomization and spraying as shown in Figure 2.7.[29] In this kind of technique, all the features are desired from the standpoint of the sensorial and textural characteristics of final products. At the beginning of the process, core material containing active molecules is dispersed in a solvent of wall material, and eventually, formation of droplets with few microns of diameter is obtained. Various newly designed nozzles are used in spray-drying-based encapsulation techniques for proper mixing of oil and carrier material to avoid the preparation of emulsion prior to atomization.[30] Further, the droplets are subjected to dehydration

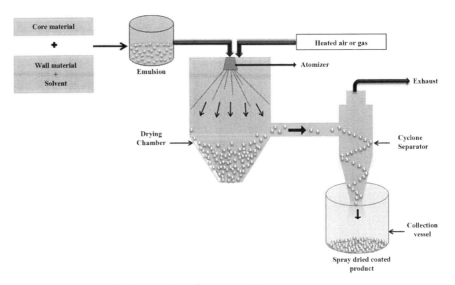

FIGURE 2.7   Spray-drying technique for encapsulation.

into capsules with diameters of 10–300 µm, in a heated chamber of the spray drier. Hot air is utilized (160–220°C) for drying the sampleswhich lasta for a few milliseconds to seconds.[31] Due to the less exposure time, heat-sensitive compounds can also be dried without characteristic loss such as flavor, aroma, taste, and so on. This process is mainly utilized for the encapsulations of hydrophobic bioactive compounds, and the process can also be initiated by dissolving the bioactive molecules in organic solutions. However, this specific technique is used rarely due to various factors such as environmental issues, safety, and cost-effectiveness.[28,32] Moreover, the effective spray-dried microencapsulates must have the capacity to encapsulate materials with a minimum oil on the surface and having a capacity of retaining maximum bioactive components.[33]

Additionally, in the spray-drying technique, one of the most important substances is the carrier materials, which should have some unique properties such as (1) good emulsification property; (2) should form a stable emulsion, (3) exhibit proper dissolution properties, (4) capable to form network, and (5) capable to form low-viscosity solutions at high concentrations.[29] Traditionally, gums including gum arabic, guar gum, and proteins such as sodium caseinate, whey protein, gelatin have been used for developing encapsulates with the aid of spray-drying technique.[29,34] In recent days, maltodextrin, mixtures of surface-active biopolymers such as gums, proteins, and modified starches with maltodextrin and other polysaccharides such as alginate have been found to provide improvement in the characteristic properties of formulated spray-dried products.[35] Moreover, surface-active agents are also used to improve the emulsifying capacity of maltodextrin, though it is having many advantageous properties including low cost, neutral aroma, taste, low viscosity, protection against oxidation, and so on.[29] The factors which predominantly influence the efficiency of encapsulation are mass ratio of core and wall material, inlet air temperature in a spray dryer. There is a contrary to wall/core mass ratio as few studies report that less core material helps to improve the efficiency of encapsulation[36] and the opposite results have also been reported.[37] A similar case has been observed when inlet gas temperature increases, which allows effective evaporation of volatiles like ethanol and water from the surface of the particles, and eventually the quantity of solid bioactive content has found to improve, which contributes to the yield of product.[38] On the other way, very high temperature during spray drying can also be responsible for the loss of volatile active compounds, which may contribute to lower efficiency of the process.[37] However, the rate of evaporation and particle formation is affected due to the higher inlet air temperature, which has an impact on the reduction in encapsulation efficiency.[39] Actually, the inlet and outlet temperatures are interrelated and the active compounds subjected to encapsulation retain for more time inside the chamber than at the outlet. One of the most common products in pharmaceuticals which are formulated by spray-drying technique is microcapsules and they are matrix type (Fig. 2.2c). In this process, usually, core materials are equally distributed throughout the wall material and the wall materials are well compacted without any cracks. This compactness of the wall material ensures the better protective nature of the encapsulated product through a reduction in gas permeability and well protection of the bioactive compounds inside. However, nonuniformity in the size and shape of the obtained particles and their tendency to form agglomerates are the limitations of this process. Yang et al. has reported about unevenness on the surface of microcapsules which is due to the rapid evaporation of liquid droplets while coming out from the atomizer.[40]

Further, this technique is responsible for various properties including decreased hygroscopicity and increased glass transition temperature, which has an influence on the stability of some compounds like protein hydrolysates.41 These properties are significantly influenced by the molecular weight and hydrophilicity of the wall materials.[40] The presence of active compounds at the surface is responsible for the rapid release of these particles from spray-dried microparticles. Further, the release of active compounds is easy in water-based food products when the spray-dried product is made from hydrophilic polymer. The release properties are based on the porosity of the particles and core to coating proportion. In addition, the release of active compounds also depends on the type of core material and based on the property of active compounds' retention of aroma upon spray drying also gets influenced.[42,27]

2.4.1.2   FREEZE DRYING

One of the foremost drying techniques for heat-sensitive food material is freeze drying. Generally, the principle of this process is based on dehydration which allows water molecules to sublimate directly from a solid phase to the vapor phase, where the temperature is below the freezing point of the material. This technique is expensive and time-taking process, which limits its application; however, it is very much fruitful for heat-sensitive food and development of high value-added product. For instance, bioactive compounds are commonly heat sensitive or susceptible to get degrade at high temperature in nature, so this technique would be one of the most appropriate techniques for encapsulating bioactive components. Papoutsis et al.[43] has compared the effectiveness of encapsulation using spray and freeze drying. Further, it has been reported that freeze drying is found to be more efficient for the encapsulation of lemon by-product aqueous extracts in comparison to spray-drying technique.[43] Moreover, spray freeze-drying technique can also be used for obtaining a better quality product. Pharmaceuticals industry recently has started using spray freeze drying for the preparation of pharmaceutical powders.[44] This combined method is usually occurred through spraying a solution-containing dissolved/suspended material using an atomization nozzle into a chamber-containing vapor of cryogenic liquid, commonly liquid nitrogen. The droplets became completely frozen while passing through the cold vapor zone of cryogenics due to the direct contact with the droplets followed by freeze drying. This technique is much effective for obtaining dry powder of narrow distributed with desired particle size by combining narrow particle size distribution of an extrusion device and freeze-drying process.

2.4.1.3   FLUIDIZED BED COATING

Microencapsulation using fluidized-bed coating is another technique by which the functionality of food products can be enhanced.[45] In this technique, powder particles are coated in a batch processor or in a continuous setup. This process involves suspension of particles by an air stream at a predefined temperature followed by spray coating. Several factors are responsible for evaporation of water including air flow, spraying rate, water content of the suspension, humidity of the air inlet, and temperature.[28,46] However, uncontrolled agglomeration of particles is the drawback in fluidized-bed process. The wet coated materials getting coalesced due to formation of network between particles.[27] Finally, the liquid bridges become solidify after evaporation and form agglomerates. To minimize the surface agglomeration, various factors need to be controlled including inlet temperature, atomization pressure, and coating solution concentration. Further, the coating material should contain some of the properties including adequate viscosity, thermal stability, and film-forming abilities. Usually, gums, proteins, derivatives of cellulose and starch, dextrin are utilized as coating material. Also, lipids including hydrogenated vegetable oils, fatty acids, emulsifiers, and waxes in molten state might be used as an effective coating material. Although many parameters need to be optimized including storage vessel, nozzle, atomizing air temperature, which should provide an ability to prevent lipid solidification. However, if the temperature is close to the lipid melting point, there may be a formulation of sticky and agglomerated particles. Further, pan coating also came out with the idea of producing the powders which will be more resistant to humidity.[46] For instance, aiming to improve the characteristics of encapsulates, several coatings can be applied onto a powder. Further, multi-wall morphology (as shown in Fig. 2.2e) can be obtained instead of only core/shell or reservoir type as shown in Fig. 2.2b. However, increased costs will be one of the main concerns of multicoating processing, which needs to be tailored. The application of fluidized-bed coating on bakery industry has an effective influence for encapsulating sorbic acid, lactic acid, potassium sorbate, and calcium propionate in bakery products. This technique is also useful for encapsulation of vitamins B, vitamin C, ferrous salts, and numerous minerals and for some acids, pigments, and flavors.[47] Fluidized-bed technique can be carried out at low temperature due to optimal heat and mass transport phenomena and equal temperature distribution phenomena, which is comparable to spray-drying technique.[48]

2.4.1.4   EXTRUSION SPHERONIZATION

Extrusion spheronization is one of the preferred and widely studied methods for fabricating uniformly sized spherical particles. The first step of this process is granulation, where bioactive compounds, stabilizers, and other ingredients are mixed with solvent or liquid binder like water to form a wet mass. This wet mass is subjected to extrusion through a die for obtaining cylindrical strands with uniform length and diameter and the extrusion is done using an oscillating granulator. The shape of these strands can be converted into spheres. To get the sphere-shaped structure, chopping of the strands followed by rounding the chopped segments into spherical shape on a marumerizer plate. The various shapes can also be obtained using conventional coating pan under forced air drying condition. Finally, the wet spheres are collected and dried through fluidized bed or tray drier prior to use. This technique is widely used for preparing encapsulates and has an application in pharmaceuticals industry where controlled release of the drug is needed.[49] The superiority of using this technique is the ease of operation, high throughput with reduced waste materials, and high drug loading.[51] Using this technique, high throughput can be obtained using suitable excipients and fillers.[50] This process can be utilized to obtain a spherical shaped pellet, where bioactive compound is embedded inside a solid carrier material which controls the release rate of bioactive molecules into the system. The physical properties of the spherical particles and release of bioactive molecules depend significantly on the starting materials.

2.4.1.5   CENTRIFUGAL EXTRUSION

Centrifugal extrusion is a liquid coextrusion process, where nozzles are used. These nozzles contain concentric orifice which is located on the outer circumference of a rotating cylinder. In addition, nozzles have a coaxial opening. The two fluids form a unified jet flow by centrifugal force, which occurs at the tip of the coaxial nozzle. This unified jet flow is responsible for forming the droplets. The core materials are fed through the inner opening, whereas carrier materials are through the outer opening and both the fluids must be immiscible. Generally, extrusion implies forcing a core material in a molten wall material mass through a die or a series of dies with a desired cross-section into a bath of desiccant liquid. The coating material gets harden on contacting the liquid and at a time started entrapping the active substances. Later on, these extruded filaments are separated from the liquid bath followed by drying and sizing.[52] Further, research is going on for optimizing the centrifugal extrusion process focusing on the recycling of the excess coating fluid from this technique.

2.4.1.6   SUPERCRITICAL LIQUID METHOD

Supercritical technologies have gained much interest in food industry due to the capability of using thermolabile products. This technology is also capable of focusing on particle engineering. It is also favorable for obtaining new natural powdered additives including flavorings, colorings, and antioxidants, with controlled characteristics such as uniformity in size. This method is most widely used in encapsulation process due to several favorable characteristics including low toxicity, low cost, easy removal, and nonflammability. Among all the available supercritical fluids, supercritical carbon dioxide ($scCO_2$) is most common. However, the selection of the supercritical process for encapsulation of active compounds is based on few factors including the solubility of the substrate and the polymer matrix in the supercritical fluid. This factor is one of the limitations of supercritical techniques. Few factors also need to be considered including size, shape, and structure of the desired particles, the production scale, and the processing costs.[53] For encapsulation based on the solubility of active compound the function of $scCO_2$ varies as solvent, antisolvent, solute, nebulization compound, and extractor, respectively. However, using $scCO_2$ as a solvent is restricted in some area of food products as substances including fats, vegetable oils, and vitamins are moderately soluble into it.[54] For instance, antisolvent, cosolvent techniques using $scCO_2$ are more effective. In antisolvent gas (GAS) technique, the active substance and wall material are solubilized in an organic solvent to form active solution. The antisolvent further able to saturate the active solution through the mass transfer process and is rapid for CO_2 because of its high diffusion rate, reduced density of the mixture, and increased the volume of the solution. The solvent power of a liquid is a function of its density, and thus the solubility of the solute containing active compound and wall material significantly decreases, reaching saturation, and then supersaturation, and at that point nucleation, the formation of fine particles occurs. In the literature, it has been reported that antisolvent technique has long been used for micronization of foods and their derivatives including citric acid, chocolate, fatty acids, and rosemary extracts,[54] for menthol encapsulation in beeswax[56] and for the stabilization of the volatile compounds derived from several essential oils, and so on.[55]

2.4.2 PHYSIOCHEMICAL METHODS

2.4.2.1 SPRAY COOLING

The encapsulation techniques such as spray cooling and spray chilling are similar to spray drying. The encapsulates are formed using atomization and the fed suspension is obtained, where the core material is dispersed in a liquefied coating or wall material. However, unlike spray drying, in spray cooling method, water evaporation is not required. The suspension which is a mixture of core and wall material is atomized under cooled or chilled air. This chilled or cooled air allows the wall material to solidify around the core. Moreover, spray chilling is done by using fractionated or hydrogenated vegetable oil with a melting point in the range of 32–42°C which is distinguished from spray cooling where the wall material is lipid such as vegetable oil with the melting range of 45–122°C. Both the methods are generally used for encapsulation of solid substances including vitamins, minerals, or acidulants. Further, utilizing the difference in melting point of the wall material, control release of the active compounds can also be possible.

2.4.2.2   HOT MELT COATING

Hot melt coating has been conducted through fluid-bed technology, where a molten coating is applied over a substrate. However, selection of coating material is most important. The viscosity, thermal stability, and film-forming ability of a coating material need to be taken into consideration. Additionally, hydrogenated vegetable oils including soybean, cottonseed, palm and canola, fatty acids, various emulsifiers, and waxes are the most suitable materials for hot-melt coating. The coating material should possess acceptable viscosity to be pumpable, atomizable to withstand processing temperatures, and be able to spread over the particle surface.[68] The principle of hot-melt coating is similar as described in fluidized-bed coating. In this coating method, the viscosity of the coating material including lipids and waxes should be in such a way that allows them to spread over the surface. In case of fluidized-bed "hot melt" process, crystallization requires to occur within a narrow temperature zone. Usually, it is speculated that plastic fats contain a three-dimensional network of fat crystals and held together by nonreversible and reversible bonds with the function of temperature. Another important factor is the deposition layers of coating which should not be sticky as stickiness induces agglomeration and influences fluidization characteristics of the bed.

2.4.2.3   IONIC GELATION

Ionic gelation or ionotropic gelation is another process which also used to encapsulate bioactive molecules into the food matrix. The outcome of ionic gelation technique is a hydrogel or a hydrocolloid gel particle. In this process, low molecular mass ions present in aqueous polymeric solution and these ions interact with polyelectrolytes of opposite charges at the beginning of the process and finally produce an insoluble gel.[58] The encapsulation consists of simply entrapment of active compounds inside the matrix and followed by their release through phase changes of gel which occurs in response to external stimuli.[59] Further, the release of active substances from the gel has been governed by various factors such as pH changes, mechanical attrition, enzymes, osmotic forces, and through diffusion.[60] Various processes are available for conducting ionic gelation including, extrusion, coextrusion or electrostatic deposition, and atomization processes.[60] In addition, stirring is one of the important factors, which helps to form an ionic solution from polymeric or hydrocolloid solution. The bioactives further added into this polymeric solution which is targeted to be encapsulated. The spherical gel structure has been formed when droplets immediately reach to ionic solution and those spherical shaped gels contain active dispersed compounds throughout the polysaccharide matrix. This technique is easy and one of the simple economically feasible methods, which does not require any kind of specific equipment.[61] However, this technique also consists of some limitations including heterogeneous gelation of gel particles due to the diffusion mechanism and soft-core formation due to the surface gelation which occurred before core gelation.[58] Further, addition of polymers influences the encapsulatios efficiency through minimizing the loss of active compounds during ionic gelation process. Some reported work mentioned the effectiveness of starch while encapsulation of polyphenols and nisin.[62,63] It has also been observed by Stojanovic et al. (2012) that molecular weight of polymers those used as fillers has some influence on the diffusion of active substances from gel beads.[64] It has also been found that low molecular weight substances had no effect on the porosity of hydrogel. Inulin is another polymer which showed influence on the release of bioactive in encapsulated molecule.[65] Proteins such as calcium caseinate and whey protein have shown effectivity on improvement of encapsulation efficiency of total phenols.[66] Caffeine, vitamins, and other active compounds with beneficial effects have been encapsulated using ionic gelation method and their impact on food systems have also been studied.[67] Moreover, other than food, the application of this technique is useful for pharmaceutical, probiotic, medical, and cosmetic products. This technique is much utilized for encapsulation of hydrophilic food compounds.

2.4.2.4 *SOLVENT EVAPORATION/EXTRACTION*

Another effective technique of microencapsulation of bioactive molecules is the solvent evaporation method. The active molecules are added in the polymeric matrix using single or various emulsion techniques. The solvent evaporation methods consist of several types of emulsions such as oil-in-water (O/W), water-in-oil-in-water (W/O/W), solid-in-oil-in-water (S/O/W), oil-in-oil-in-water (O/O/W), and oil in oil (O/O).[69] Polymeric supporting materials are dissolved in a volatile organic solvent in the single-emulsion method. The active compounds to be encapsulated is then spread or dissolved in a volatile organic solvent to make a suspension, an emulsion, or a solution.

Further, during the formation of emulsion under agitation, organic phase is emulsified into a scattering phase consisting of a nonsolvent of the polymers. This nonsolvent is immiscible with the organic solvent containing an appropriate amount of surface-active agent. After complete stabilization of emulsion, agitation is continued to evaporate the organic solvent. In general, the process of microencapsulation by the solvent evaporation method uses some common organic solvent including chloroform, dichloromethane, ethyl acetate, and ethyl formate.[70]

2.4.2.5   COACERVATION

Coacervation is considered as the original and true method of encapsulation. This method is the first microencapsulation technique which has been developed in the year 1950.[57] In this process, separation of a liquid phase of coating material has been done from a polymeric solution and further coating of that liquid phase around suspended core materials occurs as a uniform layer followed by solidification (Fig. 2.8). In Figure 2.8, the steps involved in developing encapsulated materials using coacervation technique has been detailed, where core materials are firstly dispersed in homogeneous polymer solutions (Fig. 2.8a), where formation of coacervate phase depends on the changes in solution condition (Fig. 2.8b); further, there is an initiation of coating of the core material with the coacervate phase and formation of coalescence occurs (Fig. 2.8c), which helps in obtaining encapsulated bioactive compounds. However, this technique has some limitations related to food ingredients especially for heat-sensitive food, flavors, vitamins, and other compounds. For instance, modified coacervation method has been developed aiming to encounter the hindrance. Flavored oil is usually coated using this method. Additionally, gelatin/gum acacia as a coating material is very common and well understood among several coating materials for microencapsulation. Several other coating materials are as follows carrageenan, gliadin, heparin/gelatin, chitosan, soy protein, polyvinyl alcohol, gelatine/carboxymethyl cellulose, β-lactoglobulin/gum acacia, and guar gum/dextran are also used in this method. Another type of coacervation involves the formation of multilayered microcapsule which occurred by multiple coacervation stages. In this case, one extra layer of wall material is added to the microcapsule at each passage, which finally helps to obtain a final shell layer of thickness up to 100 mm.

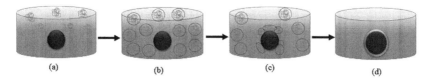

FIGURE 2.8   Coacervation technique for encapsulation representing (a) core material dispersed in homogeneous polymer solution, (b) formation of coacervate phase depends on the changes in solution condition, (c) initiation of coating of the core material with the coacervate phase and formation of coalescence, and (d) encapsulated core materials surrounded by coating material.

2.4.3   CHEMICAL METHODS

2.4.3.1   INTERFACIAL POLYCONDENSATION

Interfacial polycondensation is one of the common methods used for the preparation of heat-sensitive, infusible, stable, and meltable polymers. It is a rapid, irreversible polymerization technique. This particular polymerization reaction has occurred at the interface between an aqueous solvent and an immiscible organic solvent. The aqueous solvent contains one reactant and the organic solvent contains a complementary reactant. According to Schotten–Baumann reaction, in this process acid chlorides react with compounds containing active hydrogen atoms including –OH, –NH, and –SH.[71] This method can also be useful for emulsion diffusion, where emulsion-based encapsulation has also been done.[72] Interestingly, interfacial polycondensation is one of the promising methods used for entrapping bioactive substances for prohibiting light-induced oxidation and to obtain high encapsulation efficiency.[73] Further, the encapsulation of vitamin E, which is light, heat, and oxygen-sensitive compound, has also been done using interfacial polycondensation technique.[74] In addition, both micro and nanoencapsulation can be achieved using this technique with benefits of high encapsulation efficiency, better stability, and control of particle size.[75] This method has also been used to encapsulate bioactive such as extracted polyphenols, and so on. In the case of lipophilic polyphenols, O/W emulsion is effective for encapsulation and release of the bioactive. However, there are some limitations as O/W emulsions are usually sensitive to environmental stress including heating, chilling, extreme pH, and salt concentrations, and all of these are responsible for physical and chemical instability. Therefore, more sophisticated structures of emulsions are required for some particular applications. For instance, other forms of emulsions can also be useful.

2.4.3.2   IN-SITU POLYMERIZATION

The in-situ polymerization is one of the various available encapsulation techniques that are used for trapping active compounds. There are several studies that have been reported on this technique.[76] In this method, monomeric or oligomeric polymer solution is used as wall material and this solution is added to the core phase and eventually dispersed to the desired size. Further, the mechanism of in-situ polymerization has been detailed in Figure 2.9, where dissolution of the monomers A and B in the continuous phase in Figure 2.9a; formation of polymer (Fig. 2.9b); precipitation and deposition of polymer at the interface (Fig. 2.9c) to develop the encapsulated material has been shown. The deposition and precipitation of polymers take place at the interface due to various factors such as change in pH, temperature, or solvent quality. The solubility of monomer and polymer also affects the in-situ polymerization.[76] In the case, reactor and stirring rate are important factors to maintain the uniform size of the particles. Additionally, bioactive compounds can also be encapsulated using this technique.

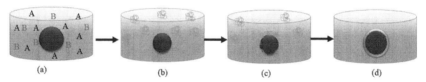

FIGURE 2.9   Mechanism of in-situ polymerization: (a) dissolution of the monomers A and B in the continuous phase, (b) formation of polymer, (c) precipitation and deposition of polymer at the interface, and (d) encapsulated core material surrounded by coating material.

2.4.3.3   INTERFACIAL POLYMERIZATION

Interfacial polymerization is one of the well-studied encapsulation techniques. In this process, reactive monomers are dissolved in two immiscible phases and the polymerization takes place at the interface, when droplet formation occurred by dispersion.[77] The mechanism of interfacial polymerization has been represented in Figure 2.10, where dissolution of monomer A and monomer B in the continuous and dispersed phase, respectively (Fig. 2.10a); diffusion of the monomers to the interface (Fig. 2.10b); polymerization reaction between the monomers are shown (Fig. 2.10c) to fabricate the encapsulated core material (active compounds surrounded by coating materials) (Fig. 2.10d). This technique is useful for various types of applications including energy storage, pharmaceuticals, cosmetics, and agriculture field and most efficiently for the development of small encapsulates. However, the reaction can be restricted if formation of a thin interfacial polymeric layer between the reagents occurred, which may also affect the mechanical property.[78] In this technique, the core phase contains reactive monomer which has a detrimental effect on encapsulated species. Further, the formation of solid microspheres rather than microcapsules can be developed if diffusion of monomers into the core phase has been occurred.[76]

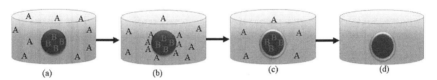

FIGURE 2.10   Mechanism of interfacial polymerization: (a) dissolution of monomer A and monomer B in the continuous and dispersed phase, respectively; (b) diffusion of the monomers to the interface; (c) polymerization reaction between the monomers; and (d) developed encapsulated core material (active compounds) surrounded by coating materials.

2.4.3.4   INTERFACIAL CROSS-LINKING

Interfacial cross-linking is another technique, where several hydrophilic polymers from different natural origins, such as gelatin, albumin, starch, dextran, hyaluronic acid, and chitosan, are solidified by the method of chemical or thermal cross-linking process. However, toxicity remains a problem for application of pharmaceutical area for using the organic solvent glutaraldehyde as a cross-linkers, although most proteins are cross-linked using this solvent. One of the examples of emulsification/interfacial polymerization is microencapsulation of lactic acid bacteria within CH membranes.[79] Further, microcapsules have been developed using the interfacial cross-linking process, where various starch derivatives are used. The resulted microcapsules can be easily lyophilized and have a characteristics feature of free-flowing nature of powders. In addition, biodegradable microcapsules are also produced, when protein is added to the starch derivative during the process.[80]

2.5 CONCLUSION

The encapsulation technique is considered as an approach to improve the delivery of bioactive compounds and living cells into food products. This is a specific kind of technique where solids, liquids, and gaseous materials are entrapped within small capsules, and the core materials can be released at controlled rates over a long period under specific conditions. The various sensitive food materials, such as essential oils, bioactive compounds, volatile compounds, can be protected from various environmental agents *via* encapsulation technique and further, the encapsulates of food materials can be incorporated in other food products to develop bioactive rich food products. The end users of the encapsulated materials include confectionary industry, convenience foods, fortified foods, functional foods, bakery industry, and so on.

KEYWORDS

- **encapsulation**
- **carrier materials**
- **core materials**
- **encapsulation techniques**
- **food waste minimization**
- **active materials**

REFERENCES

1. Shishir, M. R. I.; Xie, L.; Sun, C.; Zheng, X.; Chen, W. Advances in Micro and Nano-Encapsulation of Bioactive Compounds using Biopolymer and Lipid-Based Transporters. *Trends Food Sci. Technol.* **2018,** *78,* 34–60.
2. Augustin, M. A.; Hemar, Y. Nano- and Micro-Structured Assemblies for Encapsulation of Food Ingredients. *Chem. Society Rev.* **2009,** *38* (4), 902–912.
3. Kirby, C. Microencapsulation and Controlled Delivery of Food Ingredients. *Food Sci. Technol. Today* **1991,** *5,* 74–78.
4. Gibbs, F.; Kermasha, S.; Alli, I.; Catherine, N.; Mulligan, B. Encapsulation in the Food Industry: A Review. *Int. J. Food Sci. Nutr.* **1999,** *50* (3), 213–224.
5. Wang, W.; Jung, J.; Zhao, Y. Chitosan-Cellulose Nanocrystal Microencapsulation to Improve Encapsulation Efficiency and Stability of Entrapped Fruit Anthocyanins. *Carbohydr. Polym.* **2017,** *157,* 1246–1253.

6. Souza, J. M.; Caldas, A. L.; Tohidi, S. D.; Molina, J.; Souto, A. P.; Fangueiro, R.; Zille, A. Properties and Controlled Release of Chitosan Microencapsulated Limonene Oil. *Rev. Bras. Farmacogn.* **2014,** *24* (6), 691–698.

7. Mondal, K.; Ghosh, T.; Bhagabati, P.; Katiyar, V. Sustainable Nanostructured Materials in Food Packaging. In *Dynamics of Advanced Sustainable Nanomaterials and their Related Nanocomposites at the Bio-Nano Interface*; Elsevier: Amsterdam, Netherlands, 2019; pp 171–213.

8. Ghosh, T.; Borkotoky, S. S.; Katiyar, V. Green Composites Based on Aliphatic and Aromatic Polyester: Opportunities and Application. In *Advances in Sustainable Polymers*; Springer: Singapore, 2019; pp 249–275.

9. Wallick, D. Cellulose Polymers in Microencapsulation of Food Additives. In *Microencapsulation in the Food Industry*; Academic Press: USA, 2014; pp 181–193.

10. Heitfeld, K. A.; Guo, T.; Yang, G.; Schaefer, D. W. Temperature Responsive Hydroxypropyl Cellulose for Encapsulation. *Mater. Sci. Eng. C* **2008,** *28* (3), 374–379.

11. Kolanowski, W.; Laufenberg, G.; Kunz, B. Fish Oil Stabilisation by Microencapsulation with Modified Cellulose. *Int. J. Food Sci. Nutr.* **2004,** *55* (4), 333–343.

12. Estevinho, B. N.; Rocha, F.; Santos, L.; Alves, A. Microencapsulation with Chitosan by Spray Drying for Industry Applications—A Review. *Trends Food Sci. Technol.* **2013,** *31* (2), 138–155.

13. Farrag, Y.; Malmir, S.; Montero, B.; Rico, M.; Rodríguez-Llamazares, S.; Barral, L.; Bouza, R. Starch Edible Films Loaded with Donut-Shaped Starch Microparticles. *LWT* **2018,** *98,* 62–68.

14. Nogueira, G. F.; Fakhouri, F. M.; de Oliveira, R. A. Extraction and Characterization of Arrowroot (*Maranta arundinaceae* L.) Starch and Its Application in Edible Films. *Carbohydr. Polym.* **2018,** *186,* 64–72.

15. Ghosh, T.; Katiyar, V. Chitosan-Based Edible Coating: A Customise Practice for Food Protection. In *Advances in Sustainable Polymers*; Springer: Singapore, 2019a; pp 167–182.

16. Ghosh, T.; Katiyar, V. Cellulose-Based Hydrogel Films for Food Packaging. In *Cellular-Based Superabsorbent Hydrogels* 2019b; pp 1061–1084.

17. Ghosh, T.; Bhasney, S. M.; Katiyar, V. Blown Films Fabrication of Poly Lactic Acid Based Biocomposites: Thermomechanical and Migration Studies. *Mater. Today Commun.* **2019,** *22,* 100737.

18. Borkotoky, S. S.; Ghosh, T.; Bhagabati, P.; Katiyar, V. Poly(Lactic Acid)/Modified Gum Arabic (MG) Based Microcellular Composite Foam: Effect of MG on Foam Properties, Thermal and Crystallization Behavior. *Int. J. Biol. Macromol.* **2019,** *125,* 159–170.

19. Mohan, A.; Rajendran, S. R.; He, Q. S.; Bazinet, L.; Udenigwe, C. C. Encapsulation of Food Protein Hydrolysates and Peptides: A Review. *RSC Adv.* **2015,** *5* (97), 79270–79278.

20. Ersus, S.; Yurdagel, U. Microencapsulation of Anthocyanin Pigments of Black Carrot (*Daucus carota* L.) by Spray Drier. *J. Food Eng.* **2007,** *80* (3), 805–812.

21. Mendanha, D. V.; Ortiz, S. E. M.; Favaro-Trindade, C. S.; Mauri, A., Monterrey-Quintero, E. S.; Thomazini, M. Microencapsulation of Casein Hydrolysate by Complex Coacervation with SPI/Pectin. *Food Res. Int.* **2009,** *42* (8), 1099–1104.

22. Ortiz, S. E. M.; Mauri, A.; Monterrey-Quintero, E. S.; Trindade, M. A.; Santana, A. S.; Favaro-Trindade, C. S. Production and Properties of Casein Hydrolysate

Microencapsulated by Spray Drying with Soybean Protein Isolate. *LWT—Food Sci. Technol.* **2009,** *42* (5), 919–923.

23. Kanakdande, D.; Bhosale, R.; Singhal, R. S. Stability of Cumin Oleoresin Microencapsulated in Different Combination of Gum Arabic, Maltodextrin, and Modified Starch. *Carbohydr. Polym.* **2007,** *67* (4), 536–541.

24. Krishnan, S.; Kshirsagar, A. C.; Singhal, R. S. The Use of Gum Arabic and Modified Starch in the Microencapsulation of a Food Flavoring Agent. *Carbohydr. Polym.* **2005,** *62* (4), 309–315.

25. de Barros Fernandes, R. V.; Borges, S. V.; Botrel, D. A. Gum Arabic/Starch/ Maltodextrin/Inulin as Wall Materials on the Microencapsulation of Rosemary Essential Oil. *Carbohydr. Polym.* **2014,** *101*, 524–532.

26. Partanen, R.; Ahro, M.; Hakala, M.; Kallio, H.; Forssell, P. Microencapsulation of Caraway Extract in β-Cyclodextrin and Modified Starches. *Eur. Food Res. Technol.* **2002,** *214* (3), 242–247.

27. Zuidam, N. J.; Nedović, V. Encapsulation Technologies for Active Food Ingredients and Food Processing. *Food Science and Nutrition*; Springer: Switzerland, 2010.

28. Yanniotis, S.; Taoukis, P.; Stoforos, N. G.; Karathanos, V. T., Eds. Advances in Food Process Engineering Research and Applications. *Food Science and Nutrition*; Springer: New York, 2013.

29. Gharsallaoui, A.; Roudaut, G.; Chambin, O.; Voilley, A.; Saurel, R. Applications of Spray-Drying in Microencapsulation of Food Ingredients: An Overview. *Food Res. Int.* **2007,** *40* (9), 1107–1121.

30. Legako, J.; Dunford, N. T. Effect of Spray Nozzle Design on Fish Oil–Whey Protein Microcapsule Properties. *J. Food Sci.* **2010,** *75* (6), E394–E400.

31. Chen, Q.; McGillivray, D.; Wen, J.; Zhong, F.; Quek, S. Y. Co-encapsulation of Fish Oil with Phytosterol Esters and Limonene by Milk Proteins. *J. Food Eng.* **2013,** *117* (4), 505–512.

32. de Vos, P.; Faas, M. M.; Spasojevic, M.; Sikkema, J. Encapsulation for Preservation of Functionality and Targeted Delivery of Bioactive Food Components. *Int. Dairy J.* **2010,** *20* (4), 292–302.

33. Carneiro, H. C.; Tonon, R. V.; Grosso, C. R.; Hubinger, M. D. Encapsulation Efficiency and Oxidative Stability of Flaxseed Oil Microencapsulated by Spray Drying Using Different Combinations of Wall Materials. *J. Food Eng.* **2013,** *115* (4), 443–451.

34. Nedovic, V.; Kalusevic, A.; Manojlovic, V.; Levic, S.; Bugarski, B. An Overview of Encapsulation Technologies for Food Applications. *Procedia Food Sci.* **2011,** *1*, 1806–1815.

35. Tan, L. H.; Chan, L. W.; Heng, P. W. S. Alginate/Starch Composites as Wall Material to Achieve Microencapsulation with High Oil Loading. *J. Microencapsul.* **2009,** *26* (3), 263–271.

36. Cilek, B.; Luca, A.; Hasirci, V.; Sahin, S.; Sumnu, G. Microencapsulation of Phenolic Compounds Extracted from Sour Cherry Pomace: Effect of Formulation, Ultrasonication Time and Core to Coating Ratio. *Eur. Food Res. Technol.* **2012,** *235* (4), 587–596.

37. Rocha, G. A.; Fávaro-Trindade, C. S.; Grosso, C. R. F. Microencapsulation of Lycopene by Spray Drying: Characterization, Stability and Application of Microcapsules. *Food Bioprod. Process.* **2012,** *90* (1), 37–42.

38. Flanagan, J.; Singh, H. Microemulsions: A Potential Delivery System for Bioactives in Food. *Crit. Rev. Food Sci. Nutr.* **2006,** *46* (3), 221–237.

39. Shu, B.; Yu, W.; Zhao, Y.; Liu, X. Study on Microencapsulation of Lycopene by Spray-Drying. *J. Food Eng.* **2006,** *76* (4), 664–669.

40. Yang, S.; Mao, X. Y.; Li, F. F.; Zhang, D.; Leng, X. J.; Ren, F. Z.; Teng, G. X. The Improving Effect of Spray-Drying Encapsulation Process on the Bitter Taste and Stability of Whey Protein Hydrolysate. *Eur. Food Res. Technol.* **2012,** *235* (1), 91–97.

41. Kurozawa, L. E.; Park, K. J.; Hubinger, M. D. Effect of Maltodextrin and Gum Arabic on Water Sorption and Glass Transition Temperature of Spray Dried Chicken Meat Hydrolysate Protein. *J. Food Eng.* **2009,** *91* (2), 287–296.

42. Jafari, S. M.; Assadpoor, E.; He, Y.; Bhandari, B. Encapsulation Efficiency of Food Flavours and Oils during Spray Drying. *Drying Technol.* **2008,** *26* (7), 816–835.

43. Papoutsis, K.; Golding, J.; Vuong, Q.; Pristijono, P.; Stathopoulos, C.; Scarlett, C.; Bowyer, M. Encapsulation of Citrus By-product Extracts by Spray-Drying and Freeze-Drying Using Combinations of Maltodextrin with Soybean Protein and ι-Carrageenan. *Foods* **2018,** *7* (7), 115.

44. Costantino, H. R.; Firouzabadian, L.; Wu, C.; Carrasquillo, K. G.; Griebenow, K.; Zale, S. E.; Tracy, M. A. Protein Spray Freeze Drying. 2. Effect of Formulation Variables on Particle Size and Stability. *J. Pharm. Sci.* **2002,** *91* (2), 388–395.

45. Meiners, J. A. Fluid Bed Microencapsulation and Other Coating Methods for Food Ingredient and Nutraceutical Bioactive Compounds. In *Encapsulation Technologies and Delivery Systems for Food Ingredients and Nutraceuticals*; Woodhead Publishing: Sawston, United Kingdom, 2012; pp 151–176.

46. Onwulata, C., Ed. Encapsulated and Powdered Foods. *Food Science and Technology*; CRC Press: Boca Raton, FL, United States, 2005.

47. Dewettinck, K.; Huyghebaert, A. Fluidized Bed Coating in Food Technology. *Trends Food Sci. Technol.* **1999,** *10* (4–5), 163–168.

48. Barbosa-Cánovas, G. V.; Uliano, P. J. Adaptation of Classical Processes to New Technical Developments and Quality Requirements. *J. Food Sci.* **2004,** *69* (5), E240–E250.

49. Berner, B.; Dinh, S. Fundamental Concepts in Controlled Release. In *Treatise on Controlled Drug Delivery*; Kydonieus, A. F., Ed.; CRC Press: Boca Raton, FL, USA, 2017; pp 1–35.

50. Sinha, V. R.; Agrawal, M. K.; Kumria, R. Influence of Formulation and Excipient Variables on the Pellet Properties Prepared by Extrusion Spheronization. *Curr. Drug Deliv.* **2005,** *2* (1), 1–8.

51. Gandhi, R.; Kaul, C. L.; Panchagnula, R. Extrusion and Spheronization in the Development of Oral Controlled-Release Dosage Forms. *Pharm. Sci. Technol. Today* **1999,** *2* (4), 160–170.

52. Gouveia, L.; Empis, J. Relative Stabilities of Microalgal Carotenoids in Microalgal Extracts, Biomass and Fish Feed: Effect of Storage Conditions. *Innov. Food Sci. Emerg. Technol.* **2003,** *4* (2), 227–233.

53. Bahrami, M.; Ranjbarian, S. Production of Micro- and Nano-Composite Particles by Supercritical Carbon Dioxide. *J. Supercrit. Fluids* **2007,** *40* (2), 263–283.

54. Weidner, E. High Pressure Micronization for Food Applications. *J. Supercrit. Fluids* **2009,** *47* (3), 556–565.

55. Martín, Á.; Varona, S.; Navarrete, A.; Cocero, M. J. Encapsulation and Co-precipitation Processes with Supercritical Fluids: Applications with Essential Oils. *Open Chem. Eng. J.* **2010,** *4* (1), 31–41.

56. Zhu, L.; Lan, H.; He, B.; Hong, W.; Li, J. Encapsulation of Menthol in Beeswax by a Supercritical Fluid Technique. *Int. J. Chem. Eng.* **2010**, 1–7.

57. Srivastava, Y.; Semwal, A. D.; Sharma, G. K. Application of Various Chemical and Mechanical Microencapsulation Techniques in Food Sector—A Review. *Int. J. Food Ferment. Technol.* **2013**, *3* (1), 1.

58. Burey, P.; Bhandari, B. R.; Howes, T.; Gidley, M. J. Hydrocolloid Gel Particles: Formation, Characterization, and Application. *Crit. Rev. Food Sci. Nutr.* **2008**, *48* (5), 361–377.

59. Lakkis, J. M., Ed. *Encapsulation and Controlled Release Technologies in Food Systems.* Blackwell Publication, 2007.

60. Maestrelli, F.; Zerrouk, N.; Cirri, M.; Mennini, N.; Mura, P. Microspheres for Colonic Delivery of Ketoprofen-Hydroxypropyl-β-Cyclodextrin Complex. *Eur. J. Pharm. Sci.* **2008**, *34* (1), 1–11.

61. Comunian, T. A.; Favaro-Trindade, C. S. Microencapsulation Using Biopolymers as an Alternative to Produce Food Enhanced with Phytosterols and Omega-3 Fatty Acids: A Review. *Food Hydrocolloids* **2016**, *61*, 442–457.

62. Córdoba, A. L.; Deladino, L.; Martino, M. Effect of Starch Filler on Calcium-Alginate Hydrogels Loaded with Yerba Mate Antioxidants. *Carbohydr. Polym.* **2013**, *95* (1), 315–323.

63. Hosseini, S. M.; Hosseini, H.; Mohammadifar, M. A.; Mortazavian, A. M.; Mohammadi, A.; Khosravi-Darani, K.; et al. Incorporation of Essential Oil in Alginate Microparticles by Multiple Emulsion/Ionic Gelation Process. *Int. J. Biol. Macromol.* **2013**, *62*, 582–588.

64. Stojanovic, R.; Belscak-Cvitanovic, A.; Manojlovic, V.; Komes, D.; Nedovic, V.; Bugarski, B. Encapsulation of Thyme (*Thymus serpyllum* L.) Aqueous Extract in Calcium Alginate Beads. *J. Sci. Food Agric.* **2012**, *92* (3), 685–696.

65. Balanč, B.; Kalušević, A.; Drvenica, I.; Coelho, M. T.; Djordjević, V.; Alves, V. D.; et al. Calcium–Alginate–Inulin Microbeads as Carriers for Aqueous Carqueja Extract. *J. Food Sci.* **2016**, *81* (1), E65–E75.

66. Belščak-Cvitanović, A.; Đorđević, V.; Karlović, S.; Pavlović, V.; Komes, D.; Ježek, D.; et al. Protein-Reinforced and Chitosan-Pectin Coated Alginate Microparticles for Delivery of Flavan-3-ol Antioxidants and Caffeine from Green Tea Extract. *Food Hydrocolloids* **2015a**, *51*, 361–374.

67. Belščak-Cvitanović, A.; Komes, D.; Karlović, S.; Djaković, S.; Špoljarić, I.; Mršić, G.; Ježek, D. Improving the Controlled Delivery Formulations of Caffeine in Alginate Hydrogel Beads Combined with Pectin, Carrageenan, Chitosan and Psyllium. *Food Chem.* **2015b**, *167*, 378–386.

68. DeZarn, T. J. *Food Ingredient Encapsulation: An Overview.* 1995.

69. O'Donnell, P. B.; McGinity, J. W. Preparation of Microspheres by the Solvent Evaporation Technique. *Adv. Drug Deliv. Rev.* **1997**, *28* (1), 25–42.

70. Li, M.; Rouaud, O.; Poncelet, D. Microencapsulation by Solvent Evaporation: State of the Art for Process Engineering Approaches. *Int. J. Pharm.* **2008**, *363* (1–2), 26–39.

71. Wittbecker, E. L.; Morgan, P. W. Interfacial Polycondensation. I. *J. Polym. Sci.* **1959**, *40* (137), 289–297.

72. Janssen, L. J. J. M.; Te Nijenhuis, K. Encapsulation by Interfacial Polycondensation. I. The Capsule Production and a Model for Wall Growth. *J. Membr. Sci.* **1992**, *65* (1–2), 59–68.

73. Choi, M. J.; Soottitantawat, A.; Nuchuchua, O.; Min, S. G.; Ruktanonchai, U. Physical and Light Oxidative Properties of Eugenol Encapsulated by Molecular Inclusion and Emulsion–Diffusion Method. *Food Res. Int.* **2009,** *42* (1), 148–156.

74. Bouchemal, K. S. E. H. I. N.; Briançon, S.; Perrier, E.; Fessi, H.; Bonnet, I.; Zydowicz, N. Synthesis and Characterization of Polyurethane and Poly(Ether Urethane) Nanocapsules Using a New Technique of Interfacial Polycondensation Combined to Spontaneous Emulsification. *Int. J. Pharm.* **2004,** *269* (1), 89–100.

75. Montasser, I.; Briançon, S.; Fessi, H. The Effect of Monomers on the Formulation of Polymeric Nanocapsules Based on Polyureas and Polyamides. *Int. J. Pharm.* **2007,** *335* (1–2), 176–179.

76. Arshady, R.; George, M. H. Suspension, Dispersion, and Interfacial Polycondensation: A Methodological Survey. *Polym. Eng. Sci.* **1993,** *33* (14), 865–876.

77. Zhang, Y.; Rochefort, D. Characterisation and Applications of Microcapsules Obtained by Interfacial Polycondensation. *J. Microencapsulation* **2012,** *29* (7), 636–649.

78. Hasler, D. J.; McGhee, T. A. U.S. Patent No. 4,105,823. U.S. Patent and Trademark Office: Washington, DC, 1978.

79. Groboillot, A. F.; Champagne, C. P.; Darling, G. D.; Poncelet, D.; Neufeld, R. J. Membrane Formation by Interfacial Cross-Linking of Chitosan for Microencapsulation of *Lactococcus lactis*. *Biotechnol. Bioeng.* **1993,** *42* (10), 1157–1163.

80. Levy, M. C.; Andry, M. C. Microcapsules Prepared through Interfacial Cross-Linking of Starch Derivatives. *Int. J. Pharm.* **1990,** *62* (1), 27–35.

CHAPTER 3

The Effectiveness of Long-Term Systematic Application of Fertilizers in Crop Rotation on Sod-Podzolic Loamy Soil Low Degree of Culturality

ANATOLII A. KOVALENKO*, RAFAIL A. AFANAS'EV, and TATIANA M. ZABUGINA

Pryanishnikov All-Russian Scientific Research Institute of Agrochemistry, d. 31A, Pryanishnikova St., Moscow 127550, Russia

Corresponding author. E-mail: kovalhud@mail.ru

ABSTRACT

On sod-podzolic clay-loamy soil, with acidic and poor available forms of phosphates, the primary condition for increasing its fertility and fertilizer efficiency is the application of phosphorus fertilizers and liming. As the degree of soil culturality increased due to liming, and application of phosphorus fertilizers and manure, the order of nutrient minima changed. In first and second rotations of the seven-field crop rotation, phosphorus was in the first minimum for all crops; in the third rotation, the efficiency of certain types of fertilizers was equalized; and in the fourth rotation nitrogen and potassium came out in the first place, phosphorus took the third place.

3.1 INTRODUCTION

At present, when the problem of food security in Russia arises with certainty, an important condition for ensuring independence is a significant increase in the production and processing of crop and livestock products. In turn, the increase in the production of grain, fodder, and industrial crops is associated with the creation of favorable conditions for their growth, in particular, sufficient scientifically based use of chemical ameliorants and fertilizers. The long-term stationary experiments carried out at the Central Experimental Station of Pryanishnikov All-Russian Scientific Research Institute of Agrochemistry for several decades of the last century served to refine the fertilizer rational use for crops cultivated in the conditions of the central region of the non-Chernozem zone of the Russian Federation.

3.2   METHODS

The experiment under the index SH-1 (Shebantsevsky experimental station 1) was carried out for 30 years (from 1960 to 1989) on acidic, nutrient-poor sod-podzolic clay-loamy soil with the following agrochemical parameters in the layer 0–20 cm: pH_{salt} 4.2–4.6; acidity hydrol 4.6–4.9 mg eq./100 g; S 7.7–8.5 mg eq./100 g; humus content 1.6–1.9%; P_2O_5 by Kirsanov 12–24 mg/kg; and K_2O by Maslova 115–150 mg/kg. The program of experience with some changes was carried out in the continuation of four rotations of the seven-field crop rotation with alternating crops: vetch–oats fallow; winter wheat with clover; clover 1 year of use; winter wheat; potatoes; barley; and oat (in the first rotation—pea). Forms of fertilizers used in the experiment: semirotted manure of cattle, lime in the form of pulverized limestone, phosphorus fertilizers—phosphate powdered of Egorievsky deposit in the first and second rotations and slightly—10 kg/ha P_2O_5 in the form of granulated superphosphate when sowing grain in a row; in the third and fourth rotations—in the form of double superphosphate; nitrogen fertilizers in the form of ammonium nitrate; potassium—in the form of potassium chloride or 40% potassium salt.

Methods of application of fertilizers: manure for potatoes and vetch–oats since fall, lime since fall when fall plowed land for winter wheat and for barley, mineral fertilizers—in case of presowing cultivation for grain and in case of repeated plowing in the spring for potatoes, granulated superphosphate in the first and the second rotations in the row at planting, in the third and the fourth rotations—double superphosphate, together with other types. The size of the plots was 174 m^2 (29 × 6), experiment replication—four times. The experiment was conducted in four fields, entered sequentially, one field per year.

Varieties of cultivated crops: Vetch Lgovskaya, winter wheat PPG-186 in I rotation, Mironovskaya 808—in the second and subsequent rotations, red clover Moscow 1, potatoes Lorch, barley Wiener in I rotation, Moscow

121—in the following, oats Gamba. Technologies of cultivation of crops are generally accepted for the zone.

The scheme of the experiment included variants: without fertilizers, the background is lime (Ca), paired and triple combinations of N, P, and K on a background of lime and also NPK without Ca, two doses of manure on the background of the Ca and the dose of manure without Ca (Table 3.1). Doses of mineral fertilizers for individual crops are shown in Table 3.2.

TABLE 3.1 Experiment Scheme.

No. variants	Lime (Ca)			Manure		N		P_2O_5		K_2O	
	Per rotation (t/ha)					Annual average (kg/ha)					
	Rotations of crop rotation										
	I, II	III	IV	I, II, III	IV	I, II	III, IV	I, II	III, IV	I, II	III, IV
1	–	–	4	–		–		–		–	
2	4	2	4	–		–		–		–	
3	4	2	4	–		46.4		77.1		44.6	
4	4	2	4	–		46.4		77.1		–	
5	4	2	4	–		–		–		44.6	
6	4	2	4	–		46.4		77.1		44.6	
7	–	–	4	–		46.4		77.1		44.6	
8	–	–	4	20	40	–		–		–	
9	4	2	4	20	40	–		–		–	
10	4	2	4	20	40	–		77.1		–	

3.3 DISCUSSION

The effect of lime as one of the components of the fertilizer system on acidic sod-podzolic soils was studied in the experiment on three backgrounds: without fertilizers, against the background of manure, and against the background of complete mineral fertilizer (NPK). It was found that the effect of lime application—4 t/ha in the I, II, and IV rotation; 2 t/ha in the III rotation—significantly changed depending on the background fertilization and cultivated crop. In particular, the most stable increase in yields in the first rotation was observed only on clover (4.4–8.9 C/ha of hay) and winter wheat cultivated after clover fallow (3.1–3.5 C/ha). In the second and third rotations of the crop rotation, the effect of lime increased on all three backgrounds on clover (11.1–20.1 C/ha of hay) and remained on winter wheat after sowing in this field clover (2.8–3.6 C/ha) (Table 3.3).

TABLE 3.2 Doses of Mineral Fertilizers for Crop Rotation (kg/ha).

Doses Fertilizers (kg/ha) (a.s.)	Vetch–oats			Winter wheat			Clover	Winter wheat			Potatoes			Barley			Peas (I rotation), oats (II–IV rotation)		
							Rotations of crop rotation												
	I	II	III, IV	I	II	III, IV	I–IV	I	II	III, IV	I	II	III, IV	I	II	III, IV	I	II	III, IV
N	60	60	80	50–60	60	90	—	60	60	90	60–90	90	120	40–60	50	80	30	40	80
P_2O	30	50	60	60	60	90	—	60–70	60	90	60–90	60	120	60	50	60	60	30	60
K_2O	40	40	80	40–60	60	90	—	60	60	90	90	90	120	40–60	40	80	60	40	80

TABLE 3.3 Influence of Lime on Crop Yield and Crop Rotation Productivity (C/ha)

No. variants	Experiment variants	Vetch–oats, hay			Winter wheat			Clover 1 year of planting			Winter wheat		
		Rotations of crop rotation											
		I	II	III	I	II	III	I	II	III	I	II	III
1	Without fertilizers	11.0	15.7	27.1	6.5	6.5	14.8	23.8	12.1	31.4	11.7	12.8	13.8
2	Lime (Ca)	11.9	18.2	26.7	7.7	7.6	17.6	32.7	28.3	47.5	15.2	15.6	17.2
7	NPK	17.9	32.2	49.5	22.4	23.6	31.8	30.5	23.6	36.1	25.8	30.1	29.8
6	Ca + NPK	17.3	32.7	50.6	22.3	23.8	32.4	39.2	43.7	55.6	28.9	31.6	33.4
8	Manure 1 dose	16.4	25.1	29.9	19.3	13.2	22.9	35.9	24.9	46.2	17.7	15.9	19.8
9	Ca + manure 1 dose	16.0	26.5	37.0	19.3	15.5	23.2	40.3	36.0	58.7	20.8	19.0	22.9
P %		5.7	4.8	5.5	6.0	5.9	4.7	6.9	6.9	6.7	4.1	6.1	5.1
Grain units (C/ha)		2.9	6.4	7.5	3.6	4.2	3.4	6.7	6.1	7.0	2.8	3.8	3.6

TABLE 3.3 *(Continued)*

No. variants	Experiment variants	Potatoes			Barley			Peas—for I rotation, oats—for II and III rotation			Productivity of crop rotation for I–III rotation (C/ha, f.u.)		
		Rotations of crop rotation											
		I	II	III	I	II	III	I	II	III	I	II	III
1	Without fertilizers	103	105	90	11.9	16.1	16.3	8.5	13.0	16.1	14.9	15.8	20.3
2	Lime (Ca)	103	114	97	11.9	17.1	17.0	8.9	14.0	16.9	16.7	18.6	23.2
7	NPK	137	188	178	18.6	26.5	30.5	11.8	18.3	24.3	25.6	31.6	37.7
6	Ca + NPK	138	181	179	17.6	26.5	30.3	11.5	19.4	26.0	26.8	32.9	39.7
8	Manure 1 dose	143	172	131	14.3	21.1	19.3	12.0	15.1	19.5	23.1	23.2	27.7
9	Ca + manure 1 dose	143	179	146	14.6	23.7	22.2	12.1	16.6	20.3	24.1	26.5	31.6
	P %	5.4	4.1	4.9	4.9	4.1	3.9	5.7	3.4	4.7			
	Grain units (C/ha)	17.0	23.3	26.0	2.6	3.1	3.4	2.1	2.3	3.1			

In general, for the rotation of crop rotation, the increase in productivity from soil liming increased against the background of no fertilizers from 1.8 C/ha feed units (f.e.) in the first rotation to 2.8 C/ha f.e. and 2.9 C/ha f.e. in the second and third rotations; on a mineral background from 1.2 t/ha f.e. in the first rotation to 1.3 kg/ha f.e. and 2.0 kg/ha f.e. in the second and third rotations. Against the background of manure, the effect of lime also increased from the first rotation to the next: the increase was, respectively, 1.0, 3.3, and 3.9 C/ha f.e.

The effectiveness of manure was studied in three variants: 40 t/ha per rotation (5.7 t/ha on average per year) on limed and unlimed backgrounds and 80 t/ha—only against lime (Table 3.4). On poorly cultivated soil, manure has proved to be an effective fertilizer, increasing the yield of all crops in direct action and aftereffect. The increase from manure in a dose 5.7 t/ha averaged over the year in the first rotation on an unlimed background was 8.2 C/ha, in the second and third rotations was 7.4 C/ha. Sufficiently high increases were obtained on the calcareous background: from the first rotation to the third, respectively, 7.2, 7.9, and 8.4 C/ha f.e. In the fourth rotation, when the single dose of manure was doubled (11.4 t/ha), the increase was 9.0 C/ha f.e. Slightly smaller manure increases provided against the background of complete mineral fertilizer were provided—3.2, 4.5, 3.0, and 6.0 C/ha f.e. Double dose of manure (11.4 t/ha on average for the year in I–III rotations and 22.8 t/ha—in IV rotation) gave higher increases in crop yield and crop rotation productivity as a whole compared to a single dose. Increased productivity on a background of lime was on the rotations: 11.9, 13.4, 19.3, and 32.9 kg/ha f.e. The efficiency of pure mineral fertilizer systems was studied in five variants (according to the Wagner scheme) on a limed background and in one version without lime. In the scheme of the experiment, there were paired and triple combinations of three types of fertilizers (NPK) and, thus, the efficiency of each nutrient element was determined against the background of the other two elements. It was found that on poorly cultivated sod-podzolic soil, which was poor in the content of assimilable forms of soil phosphates (2–3 mg/100 g P_2O_5), in the first minimum for all field crops is phosphorus, not nitrogen.[1-3] Especially, it was clearly shown on grain crops (Table 3.5). Thus, at a grain yield of winter wheat on average for years of the first rotation on the limed background without fertilizers 7.7 C/ha (after vetch–oats) and 15.2 C/ha (after clover) increases from application of fertilizers directly under culture in doses of N 50–60, P 60–70, K 40–60 made at combinations of fertilizers under wheat 1 (after vetch–oats) NP—13.1 C/ha; NK—2.2 C/ha; PK—9.7 C/ha; NPK—14.6 C/ha; for wheat 2 (after clover) —respectively: from

NP—10.7 C/ha; NK—5.1 C/ha; PK—9.5 C/ha; NPK—13.7 C/ha. Thus, the yield increase of winter wheat going in vetch–oats and clover, from certain types of fertilizers (against the background of the other two species) were, respectively, from nitrogen, 4.9 and 4.2 kg/ha, from phosphorus of 12.4 and 8.6 kg/ha, and from potassium 1.5 and 3.0 kg/ha. Therefore, winter wheat after vetch–oat mixture responded strongly to nitrogen and phosphate and weakly to potash than wheat after clover. A similar effect of fertilizers was observed for barley. While yield in the control to 11.9 t/ha increase from fertilizers was at doses of 40 N–60, P, 10 + 50, K 40–60: from NP—3.5 t/ha, NK—2.2 t/ha, PK and 4.8 t/ha from NPK—5.7 C/ha, or from N— 0.9 t/ha; P—3.5 kg/ha; K—2.2 C/ha. The similar nature of fertilizer action was maintained during the second rotation of the crop rotation. On average, for the crop rotation, the increase in productivity from phosphorus at a dose of 40–46 kg/ha was 6.9 C/ha in the first rotation and 10.4 C/ha in the second. Nitrogen took second place. The increase from nitrogen fertilizer in the dose of 43–50 kg/ha was 2.9 C/ha f.e. in the first rotation of the crop rotation and 5.7 C/ha f.e. in the second rotation. The action of potassium in a dose of 47 kg/ha in most years appeared weaker and amounted to an average of 2.1–3.4 kg/ha of f.e. In general, for spring crops (barley, pea, vetch–oat mixture, potatoes), the efficiency of fertilizers was insignificant. As a result, for the first rotation of the crop rotation, the average annual productivity of 1 ha of arable land under the complete fertilizer system (N43 P46 K47) increased relative to the control only from 14.9 to 26.8 C/ha in the second rotation (N50 P46 K47)—slightly higher: from 15.8 to 32.9 C/ha f.e.

At the same time, experience has shown that a significant increase in the content of assimilable forms of phosphates in these soils is a primary condition for increasing their fertility and obtaining in the future high and more stable yields of all crops of field crop rotation.[1] It should be emphasized that in this experiment in the I and II rotations of the crop rotation, the main form of phosphate fertilizer was phosphate rock (90%) and only 10% P_2O_5 of the total amount of phosphate was applied in the form of granular superphosphate when sowing grain in a row. In III and IV rotations of crop rotation, fertilizer rates for nitrogen and phosphorus were applied by an average of 1.5 times. It should be borne in mind that the content of mobile phosphorus after two rotations of crop rotation in variants with the use of phosphorus fertilizers increased to 6–8 mg per 100 g of soil. In this regard, as well as with the increase in the rates of phosphorus fertilizer, the efficiency of nitrogen and potassium fertilizers has significantly increased (Table 3.5).

TABLE 3.4 The Effect of Manure on Yield of Crops and Productivity of Crop Rotation (kg/ha).

No. variants	Experiment variants	Vetch–oats, hay				Winter wheat 1				Clover, hay				Winter wheat 2			
						Rotations of crop rotations											
		I	II	III	IV	I	II	III	IV	I	II	III	IV	I	II	III	IV
1	Without fertilizers	11.0	15.7	27.1	26.6	6.5	6.5	14.8	18.3	23.8	12.1	31.4	36.1	11.7	12.8	13.8	20.1
8	Manure 1 dose	16.4	25.1	29.9	37.5	19.3	13.2	22.9	25.3	35.9	24.9	46.2	45.5	17.7	15.9	12.8	27.6
2	Lime (Ca)	11.9	18.2	26.7	28.4	7.7	7.6	17.6	21.3	32.7	28.3	47.5	41.4	15.2	15.6	17.2	22.1
9	Ca + manure 1 dose	16.0	26.5	37.0	36.8	19.3	15.5	23.2	27.0	40.3	36.0	58.7	50.2	20.8	19.0	22.9	29.1
10	Ca + manure 2 doses	21.2	33.6	54.7	58.9	25.9	21.1	33.0	46.1	46.8	39.4	63.9	73.9	24.0	24.3	33.9	48.0
	P %	5.7	4.8	5.5	6.1	6.0	5.9	4.7	2.9	6.9	6.9	6.7	4.5	4.1	6.1	5.1	3.4
	Grain units, C/ha	2.9	6.4	7.5	8.8	3.6	4.2	3.4	3.3	6.7	6.1	7.0	8.4	2.8	3.8	3.6	4.1

TABLE 3.4 *(Continued)*

No. variants	Experiment variants	Potatoes				Barley				Peas—for I rotation, oats—for II and III rotation				Productivity of crop rotation for I–III rotation (C/ha, f.u.)			
		Rotations of crop rotation															
		I	II	III	IV	I	II	III	IV	I	II	III	IV	I	II	III	IV
1	Without fertilizers	103	105	90	110	11.9	16.1	16.3	19.2	8.5	13.0	16.1	16.1	14.9	15.8	20.3	23.3
8	Manure 1 dose	143	172	131	200	14.3	21.1	19.3	27.9	12.0	15.1	19.5	18.6	23.1	23.2	27.7	32.1
2	Lime (Ca)	103	114	97	120	11.9	17.1	17.0	20.2	8.9	14.4	16.9	16.7	16.7	18.6	23.2	25.6
9	Ca + manure 1 dose	143	179	140	211	14.6	23.7	22.2	29.3	12.1	16.6	20.3	19.6	24.1	26.5	31.6	34.6
10	Ca + manure 2 doses	168	214	202	362	16.7	27.1	30.1	44.9	12.2	18.6	24.5	37.4	28.6	32.0	42.5	58.5
	P %	5.4	4.1	4.9	4.7	4.9	4.1	3.9	2.9	5.7	3.4	4.7	3.9				
	Grain units (C/ha)	17.0	23.3	26.0	30.1	2.6	3.1	3.4	3.2	2.1	2.3	3.1	4.3				

TABLE 3.5 The Yield of Crops and Productivity of Crop Rotation on the Variants with Mineral Fertilization System, on Average Per Rotation of Crop Rotation (C/ha).

No. variants	Doses of fertilizers, on average per year	Vetch–oats, hay		Winter wheat 1		Clover 1 year of planting, hay		Winter wheat 2		Potatoes		Peas—for I rotation, oats—for II rotation		Barley		Productivity of crop rotation (C/ha, f.u.)	
		I	II	I	II	I	II	I	II	I	II	I	II	I	II	I	II
						Rotations of crop rotation											
2	Lime (Ca) 4 t per rotation	11.9	18.7	7.7	7.6	32.7	28.3	15.2	15.6	103	114	8.9	14.4	11.9	17.1	16.7	18.6
3	Ca + N46 Ca + P45	16.2	29.7	20.8	17.1	38.9	36. 2	35.9	26.2	129	179	10.6	18.5	15.4	22.1	24.7	29.5
4	Ca + N46 Ca + K47	16.8	18.8	9.9	9.9	36.2	29.7	20.3	19.1	118	132	8.9	16.9	14.1	21.8	19.9	22.5
5	Ca + P45 Ca + K47	15.4	25.2	17.4	18.7	39.7	39.7	24.7	27.7	124	157	12.4	15.7	16.7	21.2	23.9	27.6
6	Ca + N46 P45 K47	17.3	32.7	22.3	23.8	39.2	43.7	28.9	31.6	138	181	11.5	19.4	17.6	26.5	26.8	32.9
P %		5.7	4.8	6.0	5.9	6.9	6.9	4.1	6.1	5.4	4.1	5.7	3.4	4.4	4.1		
Grain units (C/ha)		2.9	6.4	3.6	4.2	6.7	6.1	2.8	3.8	17.0	23.3	2.1	2.3	2.6	3.1		

TABLE 3.5 *(Continued)*

No. variants	Doses of fertilizers, on average per year	Rotations of crop rotation													
		Vetch–oats, hay		Winter wheat 1		Clover 1 year of planting, hay		Winter wheat 2		Potatoes		Barley		Oats		Productivity of crop rotation (C/ha, f.u.)	
		III	IV	III	IV	III	IV	III	IV	III	IV	III	IV	III	IV	III	IV
		4 years	3 years	5 years	3 years	4 years	3 years	3 years	3 years	4 years	3 years	5 years	3 years	3 years	3 years		
2	Lime (Ca) 4 t per III rotation and 8 t per IV rotation	26.7	28.4	17.6	21.3	47.5	41.4	17.2	22.1	97	120	17.0	20.2	16.9	16.7	23.2	25.6
3	Ca + N77 Ca + P69	47.3	39.3	26.4	30.5	48.3	50.6	25.5	30.8	125	198	24.2	32.2	22.0	24.0	32.3	37.3
4	Ca + N77 Ca + K77	39.9	41.2	23.9	31.8	49.2	54.1	23.7	31.2	130	221	23.8	31.6	18.8	25.5	30.2	39.0
5	Ca + P69 Ca + K77	33.2	36.3	22.9	30.4	61.8	61.4	29.1	30.9	135	207	20.5	29.9	19.5	22.3	30.9	37.1
6	Ca + N77 P69 K77	50.6	54.0	32.4	42.4	55.6	69.8	33.4	42.6	179	282	30.3	43.9	26.0	33.2	39.7	51.5
P %		5.5	5.1	4.7	2.9	5.7	4.5	5.1	3.4	4.9	4.7	3.9	2.9	4.7	3.9		
Grain units (C/ha)		7.5	8.8	3.4	3.3	7.0	8.4	3.6	4.1	26.0	30.0	3.4	3.2	3.1	4.3		

The crop yield and crop rotation productivity increased significantly for the third rotation compared to the second in lime background, in particular, for vetch–oats mixture, winter wheat, clover. Similarly, the yield of crops without phosphorus fertilizers was increased when the nitrogen–potassium fertilizer system was used. This fact should probably be attributed to the impact on the soil of lime, which helps to reduce acidity and improves the availability of natural phosphates. Productivity of crop rotation, expressed in feed units, in the third rotation on limed background was 23.2 C/ha, for variant N77P69—32.3 C/ha; for N77K77—30.2 C/ha; for P69K77—30.9 C/ha; at full mineral system N77P69K77—39.7 C/ha. Increase from certain types of fertilizers differed slightly: from nitrogen was obtained 8.8 C/ha f.e., from phosphorus—9.5 C/ha f.e., and from potassium—7.2 C/ha f.e. In the fourth rotation, compared to the third, there was an increase in the yield of winter wheat, potatoes, barley on the background of lime, as well as winter wheat, potatoes, barley, and oats when using mineral fertilizer systems. The collection of feed units per 1 ha was on the background of lime—25.6 C/ha, in the variant NP—37.3 C/ha, in the variant NK—39.0 C/ha, in the variant PK—37.1 C/ha, in the variant NPK—51.5 C/ha. The effects of individual fertilizers were also higher than in the third rotation: nitrogen yield in feed units was 14.4 C/ha, phosphorus—12.5 C/ha, potassium—14.2 C/ha; total mineral fertilizers on the background of lime—25.9 C/ha, on the background of lime + manure—22.9 C/ha.

3.4 CONCLUSIONS

1. On sod–podzolic heavy loamy soil which has poor digestible phosphates, phosphorus was the first pronounced minimum for all crops of field rotation. On average, during the crop rotation, the increase from phosphorus fertilizer at a dose of 44–46 kg/ha P_2O_5 was 6.9 C/ha f.u. in the first rotation and 10.4—in the second. To increase the fertility of such soils, it is necessary to apply phosphorus fertilizers first.

2. Liming is a necessary condition for increasing the efficiency of fertilizers on acidic poorly cultivated soils. The most responsive to liming were clover and winter wheat after clover. Highly effective fertilizer proved to be manure, which increases the yield of all crops in direct action and aftereffect both on limed soil and on the background of lime.

3. In the process of increasing of culturality soil due to liming, use of manure and phosphate fertilizer from the first rotation of seven-field crop rotation to a fourth the authors increased change of order of minimum plant nutrients and showed efficiency of the applied agro-chemicals. If in the first minimum of the first rotation is phosphorus, then in the first minimum of the fourth rotation was nitrogen, then potassium and phosphorus.

4. Additions from fertilizers on average for the crop rotation were from nitrogen (average dose 77 kg/ha) in the third rotation—8.8 C/ha f.e., in the fourth—14.4 C/ha f.e., from phosphorus (69 kg/ha), respectively, 9.5 and 12.5 C/ha f.e, from potassium (77 kg/ha)—7.2 and 14.2 C/ha f.e.

KEYWORDS

- **crop rotation**
- **crop rotation productivity**
- **soil culturality**
- **fertilizer application system**
- **fertilizer efficiency**
- **long-term experiment**

REFERENCES

1. Sigarkin, S. S.; Kuznetsova, Z. A. Study of the Fertilization System of Field Crop Rotation on Sod-Podzolic Heavy Loam Drained Strongly Acidic Soil. In *The Effect of Long-Term Application of Fertilizers on Soil Fertility and Crop Rotation Productivity*; Moscow, "Ear", 1973; pp 5–42 (in Russian).

2. Kuznetsova, Z. A.; Fetisova, N. F. Influence of Different Fertilizer Systems on Crop Yield of Field Crop Rotation and Fertility of Sod-Podzolic Cultivated Soil. In *The Effect of Long-Term Application of Fertilizers on Soil Fertility and Crop Rotation Productivity*; Kolos, M., Ed.; 1980; pp. 106–126 (in Russian).

3. Kuznetsova, Z. A.; Covic, A. D.; Fetisova, N. F. The Influence of Mineral Fertilizers, Manure and Lime under Long-Term Use in the Harvest of Crops Crop Rotation Rota, Quality, and Fertility of Sod-Podzolic Soil. *Agrochemistry* 1984, *10*, 32–41 (in Russian).

CHAPTER 4

The Effectiveness of Long-Term Systematic Application of Fertilizers in Crop Rotation on Sod-Podzolic Loamy Soil Medium and High Degree of Culturality

ANATOLII A. KOVALENKO*, RAFAIL A. AFANAS'EV, and
TATIANA M. ZABUGINA

*Pryanishnikov All-Russian Scientific Research Institute of Agrochemistry.
d. 31A. Pryanishnikova St., Moscow 127550, Russia*

Corresponding author. E-mail: kovalhud@mail.ru

ABSTRACT

On sod-podzolic medium-loamy soil of medium degree of cultivation (60–80 mg/kg P_2O_5) under favorable weather conditions, mineral and organomineral fertilizer systems provided a sufficiently high yield of potatoes, winter wheat, and perennial legumes and grasses. During crop rotation, organomineral systems were not inferior in productivity to NPK-equivalent mineral fertilizer systems. In a crop rotation with two fields of perennial legumes and two fields of potatoes, potassium was most effective, followed by phosphorus and nitrogen. On highly cultivated sod-podzolic soil (150–200 mg/kg P_2O_5 and more), the yield of winter wheat, potatoes, barley, hay vetch–oat mixture, and, in general, the productivity of crop rotation increased under the influence of nitrogen–potassium fertilizer. The effect of phosphorus fertilizer and manure was negligible.

4.1 INTRODUCTION

In the 70–80 years of the last century, as a result of the ongoing chemicalization of agriculture, the receipt of significant amounts of ameliorants, fertilizers, and plant protection products, the agrochemical science of the country was tasked to develop recommendations for the most rational use of chemicalization products. At the same time, along with the increase in crop yields, the task of radically increasing the fertility of lands, especially in the nonchernozem zone of the Russian Federation, was set. To solve the problems of effective use of chemicals, in particular, on sod-podzolic loamy soils of nonchernozem region to obtain high and stable yields of the main field crops and improve soil fertility. Studies were conducted at the Central Experimental Station of Pryanishnikov All-Russian Scientific Research Institute of Agrochemistry in long-term stationary experiments on soil of different degrees of cultivation. The results of these studies, in our opinion, have not lost relevance at the present time.

4.2 METHODS

In continuation of the studies conducted at the Central Experimental Station on poorly cultivated soil (experiment 1), field experiments were carried out on medium and highly cultivated soil. Long-term stationary field experiment on the soil of the medium level of cultivation (experiment 2) was carried out during four rotations of the seven-field crop rotation since 1970.[1] The initial soil was characterized by the following agrochemical indicators: pH_{salt}—4.1–4.3; $acidity_{hydrol}$—4.3–5.6 mg eq./100 g; $acidity_{exchange}$—0.26–0.33 mg eq./100 g; sum of exchange hydroxides—10–18 mg eq./100 g; humus content 1.6–1.8%; P_2O_5 by Kirsanov—57–80 mg/kg; and K_2O by Maslova—115–150 mg/kg. Field experiment on highly cultivated soil (experiment 2) was conducted in 1983–1990 during two rotations of the four-year crop rotation. Agrochemical characteristics of soil: pH_{salt}—5.5–6.5; $acidity_{hydrol}$—1.0–2.0 mg eq./100 g; $acidity_{exchange}$—0.05–0.10 mg eq./100 g; sum of exchange hydroxides 14–20 mg eq./100 g; P_2O_5—150–200 mg/kg; K_2O—130–200 mg/kg; and humus—1.5–1.8%.[2-4] In experiments, mineral and organomineral systems of fertilizer in field grain–grass–row crop rotations were studied. The field crop rotation on the medium-cultivated soil included the following alternation of crops: medium-late potatoes, barley with sowing of perennial grasses,

grasses of one and two years of use, winter wheat 1, early potatoes, and winter wheat 2. The alternation of crops in the field rotation on highly cultivated soil was as follows: vetch–oat mixture, winter wheat, potatoes, and barley. In the experiment on medium-cultivated soil, a variety of crops were used: potatoes mid-late Lorch and Table 19, potatoes early Priekulsky and Domodedovo, winter wheat Mironovskaya-808 and Rainbow, barley Moscow 121 and Triumph, clover Moscow-1 and Carat. In the experiment on highly cultivated soil, we cultivated the varieties of crops: winter wheat Mironovskaya-808 and Rainbow, potatoes Lorch and Istrinsky, barley Moscow-121, vetch Igovskaya, oats Eagle.

The following fertilizers were used in the experiments: cattle manure, ammonium nitrate, granulated double superphosphate, potassium chloride, lime in the form of limestone flour. In experiment 2 on medium-cultivated soil, lime at a dose of 8 t/ha was applied under the first crop rotation culture and potatoes in two doses: half for autumn plowing and the second part for fall or spring. Under the second culture, early potato manure was introduced under the autumn plowing and mineral fertilizers in the spring under the plowing of the frost. For winter wheat, phosphorus, potassium fertilizers, and a part of nitrogen (30 kg/ha) were applied in the autumn under the plow, the other part of nitrogen in the spring. Under the perennial grasses of the first year of use, phosphorus and potash fertilizers were applied after harvesting the cover barley; under the perennial grasses of the second year of use, nitrogen and potash fertilizers were applied in the spring. Under barley, all fertilizers were applied in the spring for the cultivation of the fall-plowed land.

Before putting up experiment 3, the soil was limed. In the experiment for vetch–oats, barley and potatoes, mineral fertilizers were applied in the spring for cultivation or plowing of the fall-plowed land (for potatoes). For winter wheat, phosphorus and potassium fertilizers, as well as a part of the dose of nitrogen fertilizer (45 kg/ha), were applied in the autumn; another part of nitrogen in the spring after the snowfall.

In experiment 2 on medium-cultivated soil, the mineral fertilizer system (variant 8), named conditionally basic and calculated by the balance of nutrients for the given yields: winter wheat—40 C/ha, barley—30, early potatoes—200, potatoes mid-late—250, and hay perennial grasses—60 C/ha. Other systems differed from the main system by decreasing or increasing doses of nutrients (variants 2, 11), up to the complete exclusion of one of them (pair combinations, variants 20–22). There were also organomineral

TABLE 4.1 Scheme Experiment 2: The Amount of Fertilizer Applied for Two Rotation of Crop Rotation [Manure and Lime (t/ha), N, P_2O_5, K_2O (kg/ha)].

No. variants	Fertilizer systems		Amount of fertilizer for 7 years rotation							Nutrients per 1 ha of crop rotation area			
			Lime		Manure	N	P_2O_5	K_2O	Manure	N	P_2O_5	K_2O	Total
			1st rotation	2nd rotation									
1	Lime (Ca)—background		8	4.5	–	–	–	–	–	–	–	–	–
8	Mineral—basic for a given crop		8	4.5	–	652	630	860	–	93	90	123	306
20	NK		8	4.5	–	652	–	860	–	93	–	123	216
21	NP		8	4.5	–	652	630	–	–	93	90	–	183
22	PK		8	4.5	–	–	630	860	–	–	90	123	213
2	Mineral reduced compared to the main		8	4.5	–	430	450	600	–	61	64	86	211
11	Mineral enhanced compared to the main		8	4.5	–	866	784	1124	–	124	112	160	396
14	Mineral without lime (equiv. variant 8)		–	–	–	652	630	860	–	93	90	123	306
16	Organomineral (equiv. variant 8)	1st rotation	8	–	40	441	483	494	5.7	63	69	71	203
		2nd rotation	–	4.5	40	432	490	620	5.7	62	70	89	221
18	Organomineral (equiv. variant 11)	1st rotation	8	–	70	496	526	484	10.0	71	75	69	215
		2nd rotation	–	4.5	70	481	539	703	10.0	69	77	100	246

TABLE 4.2 Experiment 2: Distribution of Fertilizers by Crops in Fertilizer Systems [manure (t/ha), N, P_2O_5, K_2O (kg/ha)].

No. variants	Potatoes mid-late				Barley			Herbs 1st y.s.		Herbs 2nd y.s.		Winter wheat 1st				Potatoes early			Winter wheat 2nd		
	Manure	N	P_2O_5	K_2O	N	P_2O_5	K_2O	P_2O_5	K_2O	P_2O_5	K_2O	N	P_2O_5	K_2O	Manure	N	P_2O_5	K_2O	N	P_2O_5	K_2O
1	–	–	–	–	–	–	–	–	–	–	–	–	–	–	–	–	–	–	–	–	–
8	–	166	145	200	80	100	100	40	50	60	50	90	100	130	–	166	145	200	90	100	120
20	–	166	–	200	80	–	100	–	50	60	50	90	–	130	–	166	–	200	90	–	130
21	–	166	145	–	80	100	–	40	–	60	–	90	100	–	–	166	145	–	90	100	–
22	–	–	145	200	–	100	100	40	50	–	50	–	100	130	–	–	145	200	–	100	130
2	–	100	90	150	60	70	70	40	50	30	50	70	80	80	–	100	90	120	70	80	80
11	–	208	170	240	100	120	130	54	97	90	97	130	135	160	–	208	170	240	130	135	160
14	–	166	145	200	80	100	100	40	50	60	50	90	100	130	–	166	145	200	90	100	130
16 I	20	110	95	90	40	80	30	40	50	60	50	90	100	130	20	70	87	74	70	80	70
16 II	20	90	90	150	62	80	70	50	60	30	60	80	90	80	20	90	90	120	80	90	80
18 I	35	120	99	73	21	70	10	54	70	90	97	130	135	160	35	65	88	7	70	80	67
18 II	35	100	100	150	60	90	103	50	60	60	60	80	100	100	35	100	100	130	80	100	100

Note: Norms of mineral fertilizers in options 16 and 18 differ between rotations due to the unequal content of nutrients in manure for 1st and 2nd rotations; lime in the 1st rotation—8 t/ha, in the 2nd rotation—4.5 t/ha.

fertilizer systems, equivalent to mineral systems in doses of applied nutrients (variants 16, 18)—Tables 4.1 and 4.2.

In experiment 3, agricultural zones of high potential fertility were initially created on highly cultivated soil by the application of lime, phosphorus fertilizer, and manure. Along with the initial background of fertility (P_2O_5 15–20 mg/kg), two mineral backgrounds were created by adding 1000–2000 kg/ha of P_2O_5, respectively, and two organomineral background when applying 800 kg/ha P_2O_5 + 80 t/ha of manure and 1600 kg/ha P_2O_5 + 160 t/ha of manure. In each of the five soil agricultural backgrounds was imposed scheme experiment including mineral and organic–mineral fertilization system (Tables 4.3 and 4.4). Soil treatment and crop care system in both experiments corresponded to zonal recommendations. The experiments were laid down in four fields. The repeatability of experiments is four-fold. The size of the experimental plots in the experiment 2—120 m² (20 × 6 m) and in the experiment 3—85 m² (17 × 5 m).

4.3   DISCUSSION

The results of the experiment on medium-cultivated sod-podzolic soil (experiment 2) for two rotations of crop rotation are shown in Tables 4.4 and 4.5. In the first rotation of the four years of cultivation of medium-late potatoes (variety Table 19), two years (1972 and 1974) were unfavorable, one year (1971) was relatively favorable (average), and one year (1973) was the best. Under favorable agrometeorological conditions, potato yield reached the planned indicators—260–280 C/ha. On average, for four years, the yield on the background of lime was 123 C/ha and on the main mineral system (variant 8)—163 C/ha. Paired combinations formed the harvest: NK—139 C/ha, NP and PK—152 C/ha. Additions of crops from individual food elements were small: from phosphorus—24 C/ha and from nitrogen and potassium—11 C/ha. Organomineral systems were not inferior in yield to purely mineral ones. Lime gave a slight (average 9 C/ha) increase in the yield of potatoes.

In the second rotation of the crop rotation (Lorch variety), two years out of four (1978, 1979) can be characterized as favorable for potatoes. In these years, the yield on the background of lime without fertilizers was 206 and 207 C/ha. Fertilizers gave an increase to 100–130 C/ha. The other two years (1976 and 1977) were unfavorable. On average for four years of the second rotation on a background of lime the crop of 167 C/ha is received.

TABLE 4.3 Scheme of Experiment 3 and Doses of Fertilizer Application for Crop Rotation [Manure (t/ha), N, P_2O_5, K_2O (kg/ha)].

No. variants	The system of fertilizer, the average for the year	Vetch–oat mixture		Winter wheat	Potatoes	Barley
		1st rotation	2nd rotation			
1	Without fertilizers	–	–	–	–	–
2	N67, 72 K97	N40 K90	N60 K90	N90 K90	N80 K120	N60 K90
3	N67, 72P60K97	N40P60K90	N60P60K90	N90P60K90	N80P60K120	N60P60K90
4	N67, 72P120K97	N40P120K90	N60P120K90	N90P120K90	N80P120K120	N60P120K90
5	N100, 109P120K97	N60P120K90	N90P120K90	N135P120K90	N120P120K120	N90P120K90
6	N67, 72P60K97 +15 t manure	N40P60K90 +30 t manure	N60P60K90 +30 t manure	N90P60K90	N80P60K120 + 30 t manure	N60P60K90
7	N67 72P60K97 + 30 t manure	N40P60K90 +60 t manure	N60P60K90 +60 t manure	N90P60K90	N80P60K120 + 60 t manure	N60P60K90

Note: In the first rotation, the unit rate of nitrogen in the fertilizer system averaged 67 kg/ha per year, in the second rotation—72 kg/ha, one and a half—respectively 100 and 109 kg/ha.

TABLE 4.4 Experiment 2: Crop Yield for Two Rotations of Crop Rotation, Average for 4 Fields (C/ha).

No. variants	Potatoes mid-late		Barley		Herbs 1st y.s.		Herbs 2nd y.s.		Winter wheat 1st		Early potatoes		Winter wheat 2nd		Crop rotation productivity (kg/ha)	
	1st rot. 1971–1974 (yy)	2nd rot. 1978–1981 (yy)	1st rot. 1972–1975 (yy)	2nd rot. 1979–1982 (yy)	1st rot. 1974–1976 (yy)	2nd rot. 1980–1983 (yy)	1st rot. 1975–1977 (yy)	2nd rot 1981–1984 (yy)	1st rot. 1975–1978 (yy)	2nd rot 1982–1985 (yy)	1st rot. 1976–1979 (yy)	2nd rot. 1983–1986 (yy)	1st rot 1977–1980 (yy)	2nd rot 1984–1987 (yy)	1st rot. f.u. (C/ha)	2nd rot. g.u. (C/ha)
1	123	167	10.4	7.9	77.9	83.7	73.2	38.7	41.1	40.0	96	142	28.5	32.0	36.2	34.2
8	163	255	16.0	17.8	82.2	105.1	88.7	53.0	44.3	47.9	214	315	34.1	49.3	47.9	51.5
20	139	213	13.0	13.6	81.6	93.4	84.1	46.3	45.3	42.9	128	190	30.8	44.6	41.3	42.6
21	152	215	13.4	13.7	79.3	91.5	82.5	47.7	38.6	40.2	142	193	29.7	40.1	40.8	46.7
22	152	211	13.0	9.5	84.3	100.6	85.7	44.4	46.8	46.8	156	261	32.9	28.3	43.7	43.5
2	159	257	14.6	15.3	82.8	99.6	86.3	48.3	44.4	45.3	191	281	34.9	40.2	45.8	49.3
11	156	250	17.1	17.6	86.9	99.5	90.7	54.9	42.3	46.1	223	325	30.8	51.0	47.2	51.9
14	154	249	15.9	16.3	74.3	98.1	85.4	45.3	44.8	48.5	214	318	33.3	48.5	45.1	50.7
16	168	262	15.0	17.6	82.7	96.0	89.7	48.0	46.6	46.6	203	313	30.7	51.8	46.9	51.4
18	168	260	14.9	17.6	88.7	99.5	90.7	53.1	41.1	43.9	205	330	33.4	52.1	47.0	52.1
LSD 05	39		3.2		3.4						48		3.4			

TABLE 4.5 Experiment 2: Efficiency of Separate Types and Different Systems of Fertilizer [Addition Yields (C/ha)].

No. variants	Systems (variants) Fertilizers and the difference between them	Winter wheat 1st		Winter wheat 2nd		Barley		Perennial herbs 1st y.s.		Perennial herbs 2nd y.s.		Potatoes mid-late		Potatoes early	
		1st rot.	2nd rot	1st rot.	2nd rot	1st rot.	2nd rot	1st rot.	2nd rot	1st rot.	2nd rot	1st rot.	2nd rot	1st rot.	2nd rot.
8	Mineral basic NPK	+3.2	+7.9	+5.6	+17.3	+5.6	+9.9	+4.3	+21.4	+15.5	+14.3	+40	+88	+118	+173
	Including: N	−2.5	+1.1	+1.2	+21.0	+3.0	+9.7	+2.0	+4.5	+3.0	+8.6	+11	+44	+58	+54
	P	−1.0	+5.0	+3.3	+4.7	+3.0	+4.2	+0.6	+11.7	+4.6	+6.7	+24	+42	+86	+125
	K	+5.7	+7.7	+5.6	+9.2	+2.6	+4.1	+2.9	+13.6	+6.2	+5.3	+11	+40	+72	+122
2	Mineral reduced (var. 2–8)	0	−2.6	+0.8	−9.1	−1.4	−2.5	−0.6	−5.5	−2.4	−4.7	−4	+2	−23	−34
11	Mineral reinforced (var. 11–8)	−2.0	−1.8	−3.3	+1.7	+1.1	−0.2	+4.7	−5.6	+2.0	+1.9	+7	−5	+9	+10
14	Mineral without lime (var. 14–8)	+0.5	+0.6	−0.8	−0.8	−0.1	−1.5	−7.9	−7.0	−3.3	−7.7	−9	−6	0	+3
16	Organomineral equiv. var. 8 (var. 16–8)	+1.3	−1.3	−3.4	+2.5	−1.0	−0.2	+0.5	−9.1	+1.0	−5.0	+5	+7	−11	−2
18	Organomineral equiv. var. 11 (var. 18–11)	−1.2	−2.2	+2.6	+1.1	−2.2	0	+1.8	0	0	0	+12	+10	−9	+15

Note: The efficiency of each of the individual food elements (N, P2O5, K2O) was calculated against the background of the other two on the main fertilizer system.

The basic system of mineral fertilizer (var. 8) provided potato yield of 255 C/ha. Similar yields were obtained by other systems of complete mineral and organomineral fertilizer. Separate nutrients created close in size increase of a crop—40–44 C/ha. Lime only slightly (The yield increased by an average of 6 C/ha) increased the yield of potatoes. In the years of cultivation of early potatoes (variety Priekulsky early) in the first rotation of the crop rotation weather conditions were quite favorable. In such years (1976, 1977, 1979), the yield on the limed background without fertilizers was 93–126 C/ha. The additional yield from the total mineral fertilizer reached 123–152 C/ha. On average for four years with a crop in the control without fertilizers 96 C/ha additional yield from the main fertilizer system was 118 C/ha, while applying nitrogen we had a harvest of 58 C/ha, applying phosphorus—86 C/ha, applying potassium—72 C/ha. Organomineral system slightly (9–11 kg/ha) was inferior to the pure mineral. The effect of lime was not observed. In the most favorable years of the second rotation (1984, 1986), the yield of potatoes (variety Domodedovo) on the background of lime was 182 and 202 C/ha. Additional yields from the mineral and organomineral systems reached 200–265 C/ha. On average, for four years, the yield on the main mineral system was 315 C/ha, the additional yields—173 C/ha. Some types of fertilizers gave additional yields: N—54 C/ha, P_2O_5—125 C/ha, K_2O—122 C/ha. Lime and manure did not have a significant impact on the crop.

Winter wheat 1 (variety Mironovskaya 808, on perennial grasses) in the years of the first rotation of the crop rotation formed a fairly high yield on the background of lime without fertilizers, on average for three years (1975–1977) about 45 kg/ha. In 1978, as a result, of the death of winter wheat was sown spring. On average, for four years (due to the less high yield of spring wheat), the yield on the background of lime was 41 C/ha. The increases in yield from the use of fertilizers (sometimes due to flatness) were insignificant. when using the main mineral system—3.2 C/ha. Of the individual elements, potassium was more effective (an increase was 2.9 C/ha), phosphorus—slightly, and nitrogen—more definitely (by 2.5 C/ha) reduced the yield of these crops. In the second rotation of the crop rotation, winter wheat (variety Rainbow) was quite productive. The average yield for four years on the background without fertilizers was 40 C/ha. Increases, as in the first rotation, were small. The largest increase in the main mineral system was 7.9 C/ha. Reduced and enhanced fertilizer systems were inferior to the main one by 1.8–2.6 C/ha. Organomineral systems were less productive than their equivalent pure mineral systems. Of the individual types of fertilizers, as in the first rotation, the most effective were potash fertilizers. The increase from potassium was 7.7 C/ha, from phosphorus—about 5 C/ha, and from nitrogen—only 1.1 C/ha. Winter wheat 2 (sown after early potatoes) for the years of the first rotation was less productive than wheat after perennial grasses. The average yield for four years on limed soil was 28.5 C/ha; it increased from fertilizer to 30–35 C/ha. The main mineral system gave an increase in yield on average 5.6 C/ha. The reinforced system was excessive, the increase from it did not exceed 2.3 C/ha. Of the individual elements, the greatest increase in yield was given by potassium —4.4 C/ha and phosphorus—3.3 C/ha, the least nitrogen—1.2 C/ha. Manure and lime did not have a noticeable effect on the crop. During the second rotation winter wheat (variety Mironovskaya 808 superior and Rainbow) have formed a relatively low yield on the background without fertilizers, on average, 32 t/ha, but gave a significant increase from fertilizers to 16–20 kg/ha. Increase from the basic mineral system was 17.3 t/ha. Of the individual food elements were the most effective nitrogen, giving an increase in the average 21 t/ha, the increase of phosphorus was 4.7 kg/ha, potassium—9.2 t/ha.

The yield of barley (variety Moscow 121) during the three years of the first rotation (1972, 1973, 1974) was formed under adverse weather conditions and amounted to 9–16 C/ha. The effect of fertilizers was minimal. And only in 1975, there was a sufficiently high yield on the background of lime—17.7 C/ha and was obtained reliable additions from the action of fertilizers up to 6–10 C/ha, although the planned yield in any year was not achieved. In the second rotation of the crop rotation, the yield of barley (variety Triumph) for three years did not exceed 3.6 C/ha on the background of lime; for experiments with fertilizers—10–16 C/ha, and only in 1980 it reached almost significant values: 16 C/ha on the background of lime and 26–30 C/ha in variants with fertilizer.

On average, for four years, the yield for the background was 8 C/ha, for the main mineral system—17.8 C/ha, of the individual elements, the largest increase was obtained from nitrogen—9.7 C/ha, from phosphorus and potassium-about 4 C/ha for each of the fertilizers.

Perennial herbs (clover–timothy mixture) of the first year of use in most years were highly productive. The yield of two mowings reached 100–120 C/ha of hay. And only in two years out of eight (1974, 1981) a fairly moderate crop was obtained—50–70 C/ha. On average, for four years of the first rotation, the hay yield on the background of lime was 78 C/ha and only slightly (by 4–10 C/ha) it increased under the influence of fertilizers. Of the individual types of fertilizer, potassium had a noticeable effect (about 3 C/ha). Nitrogen applied under the previous crop (barley), slightly (2 C/ha) reduced yield and phosphorus—no effect. More significantly lime increased the yield—an average by 8 C/ha compared to the control. The organomineral system had a slight advantage over the pure mineral system. In the second rotation, the grasses formed relatively moderate yields. The average yield for four years on the background of lime was 38.7 C/ha, on the main fertilizer system—53.0 C/ha. Addition yields from certain types of fertilizer were from nitrogen 8.6 C/ha, phosphorus—6.7 C/ha, potassium—5.3 C/ha. Lime increased the yield by 7.7 C/ha. Organomineral fertilizer systems were equivalent to mineral productivity. In general, during the rotation of the crop rotation, organomineral fertilizer systems were not inferior in productivity to their equivalent mineral fertilizer systems (Table 4.4). Productivity gains from the main mineral system (var. 8) were 11.7 C/ha of feed units (f.e.) in the first rotation and 17.3 C/ha of grain units (g.u.)—in the second. At the same time, all three types of fertilizers were effective. The first place was occupied by potassium, the increase was 7.1 C/ha in the first rotation and 9.8 C/ha in the second rotation. The second place was occupied by phosphorus, respectively: 6.6 C/ha of f.u.—in the first and 8,9 C/ha of g.u. in the second rotation. Nitrogen was less effective: 4.2 C/ha f.u. and 8.0 C/ha.

Long-term field experiment 3 on highly cultivated soil was conducted mainly in favorable weather conditions. The most favorable conditions for the formation of experimental crop yields were formed in the years of the first rotation (1983–1986), less favorable—in the second rotation (1987–1990). A sufficiently high level of crop yield was formed on limed soil without fertilization. On average, during the first rotation, we received vetch–oats hay 43.7 C/ha, winter wheat—43.5 C/ha, potatoes—233 C/ha, barley—28.0 C/ha. In the second rotation of the crop rotation, the average yield of crops, except vetch–oats, was slightly lower (Table 4.6). The differences between the different systems of fertilization were small, often within the error of experiment. Nitrogen–potash fertilizer provided an increase in grain yield on average 8–12 C/ha, potatoes—70–80 C/ha, vetch–oats hay 6–12 C/ha. Additional application of phosphorus fertilizer and manure did not lead to a significant increase of addition yields. Winter wheat was the most productive in 1983, 1984, and 1988. In these years, the yield of wheat (after vetch–oats) on the background of lime without fertilizers reached 40–50 C/ha; when applying fertilizers—up to 53–60 C/ha. In other years, this yield did not exceed 40–50 C/ha. Barley in favorable years (1984, 1986, 1987, and 1989) for variants with fertilizers formed a crop of 40–50 C/ha, in less favorable years—26–35 C/ha. Potatoes in favorable weather (1984, 1986, 1987, 1988) on the background without fertilizers created a crop of 200–300 C/ha; for variants with fertilizers—up to 300–400 C/ha. In other, less favorable years, the yield did not exceed—200–260 C/ha. The yield of vetch–oat mixture in the most productive years (1986, 1987, 1990) reached the level of 60–80 C/ha, in less favorable years—30–40 C/ha. The productivity of crop rotation on average for two rotations on the background without fertilizers was 36.6 C/ha g.u. When using systems with nitrogen–potash fertilizer, it increased to 48.6 C/ha, and when using a system with full mineral fertilizer—only to 49.6 C/ha. Fertilizer systems with additional use of manure also did not increase productivity compared to pure mineral.

TABLE 4.6 Experiment 3: The Impact of Fertilizer Systems on Crop Yield for the First (1983–1986) and Second (1987–1990) Rotation of Crop Rotation, on Average, on the Background of Cultivation for Four Fields (C/ha).

No. variants	Fertilizer system, average per year	Vetch–oat mixture		Winter wheat		Potatoes		Barley	
		1st rot.	2nd rot.	1st rot.	2nd rot.	1st rot.	2nd rot.	1st rot.	2nd rot.
1	No fertilizers	43.7	45.2	43.5	31.3	233	174	28.0	25.2
2	N67, 72 K97	49.5	57.2	51.3	44.4	306	258	38.8	35.0
3	N67, 72P60 K97	50.3	58.9	51.7	44.7	312	264	39.7	35.5
4	N67, 72P120K97	51.2	58.2	51.5	44.8	316	267	40.2	35.7
5	N100, 109P120K97	50.3	60.5	49.5	47.5	319	271	37.5	37.1
6	N67, 72P60K97 + 15 t of manure	51.1	61.7	51.9	44.7	316	276	39.8	36.2
7	N67, 72P60K97 + 30 t of manure	51.5	62.6	52.5	44.6	314	284	40.5	37.0

4.4 CONCLUSIONS

1. On sod-podzolic medium-loamy soil of medium degree of cultivation (60–80 mg/kg P_2O_5) under favorable weather conditions, mineral and organomineral fertilizer systems provided a sufficiently high productivity of potatoes, winter wheat, and perennial legumes and grasses. Organomineral fertilizer systems were not inferior to equivalent doses of nutrients to mineral systems in terms of yield of potatoes, perennial grasses, but they were slightly less effective when winter wheat was cultivated.

2. On the soil of the average degree of cultivation, individual nutrients (N, P_2O_5, K_2O) created similar increases in the potato crop. On this soil, nitrogen was more effective when second-year grasses were cultivated, whose herbage consisted mainly of cereal species; lime did not have a significant effect on the yield of potatoes and winter wheat, but significantly increased the yield of perennial grasses.

3. On highly cultivated soil (150–200 mg/kg P_2O_5 or more), a fairly high yield of winter wheat, potatoes, barley, and vetch–oat mixture was formed on a limed background without fertilization. Moderate doses of nitrogen–potassium fertilizers under favorable weather conditions increased the yield of all crops in the rotation. Phosphorus fertilizer and manure (against the background of NPK) did not significantly affect the yield of field crops. Fertilizer systems with full mineral fertilizer, as well as organomineral did not have a significant advantage in the case of using these soils over mineral with nitrogen–potash fertilizer.

4. On highly cultivated soil, winter wheat (varieties Mironovskaya 808 and rainbow) after long-term grasses formed a fairly high yield (40–45 C/ha) on the background of lime without fertilizers, but winter wheat reacted poorly to fertilizers. Winter wheat, which grew in the crop rotation after potatoes, had a low yield on the background without fertilizers (on average 28–32 C/ha).

KEYWORDS

- **soil**
- **fertilizers**
- **crop rotation**
- **productivity**

REFERENCES

1. Efimova, A. S.; Zolotarev, V. P.; Balakina, N. I.; Chevzhik, V. P.; Tishchenko, A. G. Influence of Various Fertilizer Systems on the Yield of Field Crops, Crop Rotation Productivity and Fertility of Sod-Podzolic Heavy-Clay Medium-Cultivated Soil. In *Scientific Works of Agricultural Sciences: Effect of Long-Term Application of Fertilizers on Soil Fertility and Crop Rotation Productivity*. Moscow: "Ear" 1980; pp 81–105 (in Russian).

2. Mineev, V. G.; Covic, A. D.; Kovalenko, A. A.; Trofimov, S. N. Influence of Mineral and Organo-Mineral Fertilizer Systems on Yield and Quality of Crops of Field Crop Rotation on Cultivated Sod-Podzolic Soil. Message 1. Levels of Cultivation of Sod-Podzolic Soil, Plant Productivity and Fertilizer Efficiency. *Agrochemistry* **1988,** *6*, 3–12 (in Russian).

3. Kovalenko, A. A.; Vaulin, A. V.; Afanasiev, R. A. Influence of Phosphorus Fertilizers on Agrochemical Properties of Sod-Podzolic Soil and Crop Yield. *Agrochemistry* **2009,** *10*, 3–10 (in Russian).

4. Mineev, V. G.; Kovalenko, A. A.; Vaulin, A. V.; Afanasiev, R. A. Influence of Phosphate Agrophones on the Efficiency of Fertilizers and Productivity of Crops of Left Crop Rotation on Sod-Podzolic Soil. *Agrochemistry* **2009,** *11*, 22–32 (in Russian).

CHAPTER 5

Synthesis of Carbon Dots from *Citrus limon* Peel by Microwave-Assisted Process and Its Application for Detection of Ferric Ion (Fe^{3+}) and Development of pH Paper

MALABIKA KALITA, MANISHA MEDHI, MONICA YUMNAM,
ARUN KUMAR GUPTA, and POONAM MISHRA*

*Department of Food Engineering and Technology, Tezpur University,
Tezpur 784028, Assam, India*

Corresponding author. E-mail: poonam@tezu.ernet.in

ABSTRACT

Utilization of waste is becoming a good option from which numerous useful components can be extracted or synthesized. In this chapter one of such application will be discussed. *Citrus limon* peels, one of such kind of waste from which a good source of many bioactive and organic compounds can be extracted. Carbon dots are the nanomaterials which size lies below 10 nm, which can be synthesized from many natural and chemical sources. These carbon dots are rich in many properties like fluorescence, photostability, etc which can be utilized in many fields like sensing, bioimaging. So, basically in this chapter synthesis of carbon dots form low-cost raw product, methods of synthesis, and utilization of the same in the sensing purposes of ion like Fe^{3+} in water will be discussed. Also, discussion will be extended to prepare some pH-based paper by the utilization of the same for sensing of pH in both acidic and alkaline medium which can be visually distinguished.

5.1   INTRODUCTION

Carbon dots, popularly known as C-dots, are the carbon nanomaterial with 10 nm or below than that of size, and with introduction and leading researches on the same, with time it is receiving many positive responses and interests as it has many tuneable characteristics like small in size, good fluorescence giving property, and easy and cost effective way of recovery. Many organic components are becoming major sources or basis for the synthesizing C-dots. With momentum of recovery of value added products from waste materials, the same can be the base material for the synthesizing purposes of C-dots. With the unique property of fluorescence, it can be utilized in the areas like sensing, bioimaging, development of probe, drug delivery, etc.

Green chemistry is a newer approach of scientists and researches which cover the conversion of biodegradable waste into value-added products. With industrialization, the major waste percentage is coming from the wastes generated from the peels of different fruits and vegetable origin of different fruit processing industry. Presently, waste management has become an obligatory chapter for every domain. So, recycling and extracting of some value-based compounds from wastes of such origin is becoming mandatory. These wastes are used to produce C-dots by an unsophisticated and inexpensive carbonization procedure. The different sources C-dots play great role in distinct analytical chemistry and applications. With the nontoxic property and excellent sensitivity of C-dots, it is used in detection of many heavy metals, pesticides, etc which are hazardous to human life.

Iron (Fe), an essential element of the biological system, which exhibits mainly in two oxidation states +2 and +3 states which is known as ferrous, [Fe(II)] form and ferric form [Fe(III)], respectively. In the living system or organism this redox states of Fe^{3+}/Fe^{2+} has an essential role. But the role is limit dependent. With beyond or below the limit, that is, with mismatching of concentration as requirement, the cells, tissues, organ system reaches in a risky stage with forming reactive oxidative intermediates as Fe ion readily take part in oxidation–reduction reactions.[26] Moreover for the blood-related diseases to overcome these iron is an essential metal. The water is one of the main components required by living being as a life supporting medium as many complex chemical reactions are ongoing in the body with water as a medium. And interestingly, the main abundance source of iron is water, through which, it can be easily received by organism. But the point strikes when the level of iron is sometimes above range or crosses the minimum permissible limit and such type of water is when consumed by people can cause adverse effect to their health. So, determination of Fe in the trace level for staying in safer side is always important. There are many techniques for Fe^{3+} sensing such as colorimetric,[47] potentiometric,[16] fluorimetric,[13,50] flame atomic absorption spectroscopic,[1] and mass spectrometric.[34] However, these methods are sophisticated and needs lots of facilities like well equipped laboratory facility, skilled person to handle, time consuming process. So, there is a need of low-cost and less time-consuming process for the detection of the same with same level of accuracy. The C-dots fulfils the criteria as found in many literatures. In this chapter, we will discuss about the different routes, sources, properties, and applications of C-dots with basic emphasis in the area of detection of Fe^{3+} ion in water and development of pH paper-like strip with C-dots with visual differentiation of color in different pH range.

5.2   CARBON DOTS—AN EMERGING NANOMATERIAL

Origin of fluorescent C-dots was the surface defects observation in single and multiwall nanotubes. Surface passivation, another technique was utilized for enhancing the luminescence of the carbon nanotubes.[19] All the above observation led to the discovery of carbon dots. Xu et al. discovered the fluorescent C-dots during the separation and purification of single-walled carbon nanotubes (SWCNTs) and reported in his publication in 2004.[44]

Generally, less than 10 nm average size has been reported for C-dots. The lowest size enhances the use of it in biological applications. It is gaining importance as it gives fluorescence under UV light. But based on the raw materials, routes of synthesis, pH, concentration, time temperature combination of synthesis may affect this fluorescence property. Surface passivation and doping are the widely used techniques for enhancing the quantum yield (QY) and fluorescence property. The surface of C-dots can be modified with introducing different impurities like different functional groups, doping agents, etc. Generally, nitrogen, sulphur, boron, and sometimes phosphorus were used to enhance the properties of the C-dots.[3,43] Among these, the nitrogen doping was found to be more fruitful.[4] Doping tunes more prospects in the physicochemical properties of C-dots. Both the techniques increase the size of the C-dots. Addition of this is basically the idea of introducing impurities in the surface to enhance the strategic role of electron–hole pair.

C-dots has unique fluorescent properties, like strong fluorescence, good photostability, broad excitation spectra, tunable emission spectra which are distinctive features and differ from all other fluorophores and inorganic or organic dyes. In this regard, green synthesis has acclaimed role in this field of research. Green chemistry suggests the use of cost effective, eco friendly, and environmentally safe materials. In this accordance, plant-based origin materials are sound effective as a base material for approaching with green synthesis. Green synthesis has grabbed the attention and reaches in a sound level as the way and products are feasible, less toxic, highly economical, and biocompatible with characteristics low costing property as less time and energy consumption occurs.

The main elements of C-dots are C, O, and H. The compositional ratio varies with precursor used for synthesis. Different physical and chemical synthetic routes have been reported for the synthesis of C-dots, which would endow them with various characteristics properties. The strong up-conversion fluorescence of C-dots with two-photon excitation in the near-infrared region has also been recently discovered, which was favorable and significant for in vivo bio applications and photo-assisted catalytic reactions.[22,30] With emerging luminescent nature of the material, sustainable efforts have been given to improve the synthesis, to know the on lying chemistry, performance, mechanism, and applications of C-dots. Although the C-dots has many applications in various fields due to its diverse properties but sometimes application is difficult in biological model as preparation methods of C-dots adversely affect the size of the C-dots, their distribution, and hence ununiformity in photoluminescence (PL) behavior. Hence, sometimes after synthesis purification steps are needed and done with filter paper, membrane with different molecular cutoff weight. The mechanism of PL in some point is still unclear like multiphoton emission behavior. Hence, major emphasis have been given in near future to produce C-dots on a large scale with economical and green route to obtain them with good yield, high QY, uniform size, and strong fluorescent.

5.3   SYNTHESIS OF CARBON DOTS

5.3.1   FOOD: A COST EFFECTIVE SOURCE FOR SYNTHESIS OF CARBON DOTS

Different organic materials have been used for the synthesis of C-dots till date. Carbon, which is available in almost all organic materials, can be extracted via green route. Different carbon-rich chemical and synthetic components along with different natural products have been enlisted as a source of C-dots. Starting from synthetic chemicals like glucose,[46] citric acid,[12] food products like garlic,[48] peels of pomelo,[24] limon[6] up to natural products related to human like nail,[7] etc were also reported as successful source for synthesizing the C-dots.

With the demand of green, nontoxic, and environment friendly compounds, different natural compounds have been reported till date for synthesizing C-dots. Researchers found that different food-based materials are good source for synthesizing C-dots. Different kind of waste materials were also utilized till date for preparing C-dots and successfully utilized in many field of applications. Carbohydrate like monosaccharides, polysaccharides-based precursors are the main carbon source to synthesize low-cost C-dots. Carbohydrate with low molecular weight mainly glucose, sucrose, fructose, polymeric carbohydrates like chitosan, alginic acid, starch, and other polysaccharides are the largely used examples of carbon source. Raw material containing carbohydrate such as bagasse, juice of different fruits origin like grape juice, orange juice, sugar cane juice, leaves of plant like tulshi, peels of different fruits, raw food products like milk, garlic, coffee, wheat have also been studied. Low-cost viability and sustainability are some of the important criteria for selecting carbohydrates as a base material for making C-dots.

Wang and Zhou reported hydrothermal method for synthesizing C-dots from milk.[37] The C-dots were reported to be used in imaging of U87 cells which is a human brain glioma cancer cell line. Hence, they reported the N-doped C-dots from milk as excellent for optical imaging of the mentioned cell line. Another green synthesis of fluorescent C-dots was reported by Arul et al. where they used the fruit extract of *Actinidia deliciosa* popularly known as Kiwi fruit and for enhancing the fluorescence, the C-dots were doped with nitrogen and ammonia solution.[2] The reported N-doped C-dots was with average size of 3.59 nm and showed low cytotoxicity and well biocompatibility on L-929 and MCF-7 cell. Also the synthesized N-doped C-dots exhibited excellent catalytic activity in the presence of sodium borohydrate in reduction of Rhodamine-B. Fruit waste materials like pomelo peels were treated with hydrothermal method for preparing blue fluorescent giving C-dots with a Teflon lined autoclave and used for novel sensing technique of mercury (Hg^{2+}) ion in water samples with detection limit 0.23 nM.[25]

As discussed above, carbohydrate-based precursors with different composition of monosaccharides, polysaccharides, and other organic raw materials are the main platform base material as carbon source to synthesize low-cost C-dots. So, in this chapter discussion will be carried on the low cost, easily available *Citrus limon* peel for synthesis of C-dots and their applications.

5.3.2   DIFFERENT APPROACHES OF SYNTHESIZING CARBON DOTS

By variety of approaches, C-dots can be synthesized. Among them, the approaches can be broadly classified in two heads: "top-down" and "bottom-up". Under the two heads, different ways are there as shown in Figure 5.1. In the top-down methods, arc-discharge, laser-ablation, electrochemical, etc and in bottom-up methods, hydrothermal, microwave, combustion, pyrolysis, etc are included. Among them, hydrothermal and microwave-assisted methods were widely used as they are easy to handle, low cost, and less time-consuming approaches and gives C-dots with rich physicochemical properties. In the first approach, the breakdown of bulk carbon materials is done under relatively rigorous conditions.

While in the "bottom-up approach", C-dots are generally synthesized from small carbon-containing molecules via "polymerization–carbonization" process.[27]

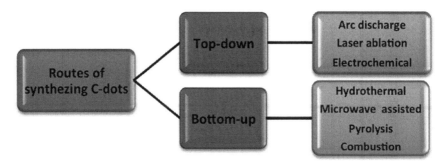

FIGURE 5.1   Diagrammatic view of different routes for synthesizing C-dots.

5.3.2.1   TOP-DOWN APPROACHES

5.3.2.1 1   Laser Ablation Method

By laser ablation of carbon in presence of water vapor and argon gas at a temperature of 900°C and at pressure 75 kPa Sun and coworkers produced

C-dots. After refluxing in HNO_3 for up to 12 h and surface passivation, C-dots gave bright luminescence emission.[36,20,41] The C-dots obtained as such were in aggregation of various size and the C-dots had no fluorescence. Then treatment were given by refluxing with an aqueous solution of nitric acid for 12 h, and then reacted with polyethylene glycol (PEG1500N) or poly(propionylethyleneimine-coethyleneimine) (PPEI-EI). The results of such treatment generate new ideas as the passivated C-dots were in size of 5 nm and were of strong photoluminescent property.[53] So with proper selection of organic solvents, the emission behavior of C-dots can be modified as luminescences and is closely related with the surface states of the C-dots.

5.3.2.1.2   Arc-Discharge Method

Xu et al., during purifying SWCNTs derived from arc-discharge soot, isolated an unknown fluorescent carbon nanomaterial.[44] First a stable suspension was obtained as oxidized the sample with HNO_3 and extracted the same in NaOH solution with pH 8.4. Then electrophoresis was run, and fluorescent C-dots were separated with average size 18 nm. The yield of C-dots obtained from arc discharge was very low. Also, arc discharge soot contained a variety of complex components, which were later on difficult to purify, obtain the pure C-dots.[53]

5.3.2.1.3   Electrochemical Methods

Zhang and coworkers proposed preparation of carbon quantum dots via the electrochemical carbonization of low-molecular weight alcohols.[10,41] As the reference electrode, a calomel electrode mounted on a freely adjustable Luggin capillary was used and two Pt sheets were used as the working and auxiliary electrode. Under basic conditions, the alcohols were transformed into CQDs after electrochemical carbonization. With increase potential, the size and graphitization of the synthesized carbon quantum dots increases. The resultant carbon quantum dots showed excellent excitation and size-dependent PL properties without employing complicated purification and passivation procedures.

5.3.2.2   BOTTOM-UP APPROACHES

5.3.2.2.1   Thermal Method

Pyrolysis and carbonization under controlled environment and at high temperature is one of the method of synthesizing low-cost C-dots. Here, the samples are reduced in size and using porcelain crucibles. Generally, the samples are burned or heated in Muffle furnace or oven. From caffeine pyrolysis via Muffle Furnace at 180°C for 3 h, blue fluorescence C-dots were prepared with QY 69%, which were used in silver ion detection in aqueous solution.[8]

5.3.2.2.2   Microwave-Assisted Method

Microwave-assisted synthesis of C-dot is the most commonly used process as it is very cost effective and easily availed process. Also, the time of synthesis is very less compared to the other methods. It generally requires less than 20 min in preparing the C-dots and gives good yield. It gives a good morphology of structures due to the use of high-pressure and high-temperature conditions during the process. Moreover, it is economically viable as the solvent used is water. For proper synthesis, the power and time combination of microwave is an essential parameter, mismatch of use gives dull fluorescence, enlarged size, and nonuniform distribution of C-dots with low stability over time. Kalita et al. stated the use of microwave method for the preparation of C-dots from waste peel of *Citrus limon* at a optimized microwave power and time combination of 900 watt (W) for 8 minutes (min). Optimization was carried out with analysis of carbon, hydrogen, and nitrogen percentage and QY of the synthesized C-dots at various time and power combinations. And found that 900 W gave highest yield with carbon 60.32%, hydrogen 9%, and nitrogen 8.2%. And 8 min synthesis time gave carbon 62%, hydrogen 9%, and nitrogen 8.2% with highest QY of 56%.[15] For enhancing the properties like fluorescence, m-aminobenzoic acid (MAC) was used. Generally, to discard the large size particles centrifugation, filtration and dialysis are employed and supernatant are collected for the study and application purposes.

Citrus limon peel   **dried peel**   **carbonization = C dots**

FIGURE 5.2   Synthesis of C-dots from *C. limon* peel.

5.3.2.2.3   Hydrothermal Method

Hydrothermal method is one of the popularize methods for preparation of C-dots. Generally in this approach, combination of parameters like pressure, temperature, and time eventually affects the synthesis of C-dots. Teflon lined autoclave are generally used for the synthesis. It is small pressure vessel in which sample solution is given and placed in the apparatus in a proper temperature condition generally within 180–250°C for 5–12 h and normally cooled down. After that the C-dots solution are further dialysis to obtain pure C-dots. 1000–3500 KDa molecular cutoff dialysis membrane are used for the purpose. Yang et al. reported the preparation of C-dots from glucose for imaging purpose.[46] Different food-based materials were used to synthesize C-dots via hydrothermal methods like pomelo peel,[24] bagasse,[17] soy milk,[51] orange juice.[33] They showed blue or green color fluorescence. The high fluorescence property is due to may be the formation of carboxyl and hydroxyl functionality onto the surface of C-dots.[45]

5.4   PROPERTIES OF CARBON DOTS

The characterization of C-dots is done by various instrumentations. Basically, for characterization purposes, ultraviolet–visible (UV–Vis) spectroscopy, fluorescence spectroscopy, X-ray diffraction (XRD), Raman spectroscopy, Fourier transform infrared spectroscopy (FTIR), CHN analyzer, scanning electron microscopy (SEM), transmission electron microscopy (TEM) are used. C-dots generally absorbed light in shorter wavelength of electromagnetic spectrum and emits light in longer wavelength. The electron that C-dots absorb has higher energy than the electron that it emits. The fluorescence behavior of C-dots is excitation and emission dependent. C-dots showed strong absorption in the UV region and as it tends to longer wavelength, absorption intensity decreases with extending a tail like structure in the visible region. The excitation and emission-dependent behavior allows different color emission of the synthesized C-dots which allows it to apply in multicolor imaging applications. The mechanism of PL behavior of C-dots can be explained in three heads: (i) due to the intrinsic bandgap in the C-dots arise from the quantum confinement effect or the conjugated π-domains, (ii) due to the surface defects or the creation of trap states by introducing impurities in the bandgap with activity like doping, surface functionalization of C-dots, (iii) due to the presence of individual fluorescent molecules (fluorophores) in the C-dots.[5,55] Based on these theories, the tunable emissions behavior of C-dots has been attributed to the broad size distribution, different color emission properties, and variable surface chemistry.

5.5   CARBON DOTS CHARACTERIZATIONS

As discussed above, many instruments are used to study the characterization of the synthesized C-dots. Detailed discussion is carried out below:

Carbon Hydrogen Nitrogen (CHN) analysis:

For determination of carbon (C), hydrogen (H), and nitrogen (N) elemental concentrations in C-dots CHN analyzer is used which is basically worked in combustion principle and with detectors the elemental analysis is done.

QY analysis:

Photostability and fluorescence are two major points that should be taken into account in the application of luminescent C-dots. The QY of C-dots were generally determined by comparing the fluorescence emission and absorbance behavior with a standard dye. Generally, dyes are used in quinine sulphate, rohadamine 6G, fluorescein. The mostly used and reported standard was quinine sulphate in 0.1 M H_2SO_4 which has QY 54% at 340 nm excitation wavelength. Same concentration series as standard dye was run for sample followed by making standard curve to obtain synthesized C-dots QY.[9] The QY of the C-dots was calculated by the following equation (De and Karak, 2013; Kalita and Mishra, 2017):

$$Q = Q_R \times \frac{I}{I_R} \times \frac{A_R}{A} \times \frac{n^2}{R_n^2}$$

Where Q is the QY, I is the measured integrated fluorescent emission intensity, n is the refractive index of the solvent, and A refers to the absorbance. The subscript R refers to the corresponding parameter of known fluorescent standard.

TABLE 5.1 Quantum Yield of Various Synthesized C-Dots from Different Sources.

Serial no.	Raw material used	Method of synthesis	Quantum yield (%)	Reference
1.	Citric acid and cysteamine	Microwave assisted	28.0	[28]
2.	*Aegle marmelos* leaves powder	Microwave assisted	22.0	[29]
3.	Caffeine	Pyrolysis in muffle furnace	38.0	[8]
4.	Lycii Fructus	Hydrothermal Teflon lined autoclave	17.2	[35]

Morphology and particle size determination by transmission electron microscope (TEM):

For determination of size distribution and morphology of the C-dots TEM are used. The samples for TEM characterization were prepared by coating the given volume of C-dots dispersion on a carbon-coated copper grid at 200 KV, and then dried under vacuum at 50°C for 12 h.[15] The obtained C-dots prepared from peel of *Citrus limon* peel appeared relatively uniform in size ranging from 1.5 nm to 3.5 nm, spherical and well dispersed.

PL property of C-dots:

PL behavior is the classic signature of C-dots and considered as most significant property from the application viewpoint. Fluorescence emission spectra of C-dots are taken generally with fluorescence spectrophotometer. In a cuvette, sample is given and excites at a particular wavelength to obtain the emission behavior of the C-dots. It involves using a beam of light, usually ultraviolet light, that excites the electrons in molecules of certain compounds and causes them to emit light; typically, but not necessarily, visible light. A complementary technique is absorption spectroscopy. Generally, the excitation peak is obtained at shorter wavelength and emission peak is mapped in longer wavelength. Stoke shift is an important criteria during analysis of the fluorescence behavior for stability and application purposes. Generally for the application purposes, the continuous scan was run to get the highest emission peak and the excitation wavelength at which the maximum emission peak was obtained used for further detection purposes.[15]

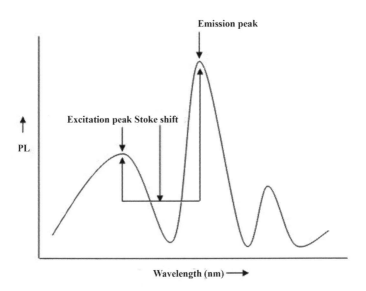

FIGURE 5.3   Diagrammatic view of fluorescence curve obtained from fluorescence spectrophotometer.

Analysis by UV–VIS:

The UV spectra of synthesized C-dots were obtained by UV–VIS spectrophotometer. This technique is complementary to fluorescence spectroscopy; in that fluorescence deals with transitions from the excited state to the ground state, while absorption measures transitions from the ground state to the excited state. Generally, the sample were scanned at 200–800 nm and obtained spectra were recorded. During the UV-V spectroscopic measurements of C-dots, an absorption band are obtained near 230–350 nm suggesting the formation $C=C$ which signifies the carbonization process which is nothing but formation of double or triple bond. In the range of wavelength from 240–270 nm and 300–320 nm, the absorption were generally R–ℜ* transition of $C=C$ and n-π* transition of carbonyl and amine functional groups, respectively.[7]

FTIR of synthesized C-dots:

Functional groups on C-dots are characterized using FTIR. It simultaneously collects spectral data in a wide spectral range. Prepared C-dots are mainly composed of C, H, O, and N. The spectral peak obtained for characteristic absorption bands of the −OH stretching vibration mode at about 3400 and 1073 cm⁻¹ could be observed. The band at 2923 cm⁻¹ corresponds to the C−H stretching mode. In addition, the peaks appearing at 1590 and 1400 cm⁻¹ may be caused by the asymmetric and symmetric stretching vibration of COO⁻, respectively. These findings provide evidence that both the hydroxyl and carboxylic groups originated from carbohydrates.[48]

Photo stability of synthesized C-dots:

Photostability is a remarkable characteristic of C-dots which is generally done by irradiating the synthesized C-dots under UV light continuously for a period of time. As reported by Kalita et al., the photostability of the C-dots was tested by irradiating the sample by a UV lamp at a wavelength of 365 nm for 6 h continuously and fluorescence intensity was measured at an interval which showed no such remarkable loss of fluorescence over increased time.[15] This phenomenon indicated the good photo stability of the synthesized C-dots.

5.6   APPLICATION OF CARBON DOTS

C-dots are rich in many properties. They have excellent PL property with high biocompatibility, low cytotoxicity, good photostability which leads broad areas of applications starting from development of sensors, drug delivery, bioimaging, catalysis, optoelectronics, etc.[14] For application in the bioimaging field optical and photostability properties are the most important properties of C-dots. Most widely use are of applied C-dots was till date in detection of different analytes of interest. Different metal ion like arsenite,[28] mercury,[31] lead,[18] iron,[29] copper,[11] silver[8] were reported by detection with C-dots synthesized from citric acid, flour, tulsi leaves, bael (*Aegle marmelos*) leaves, folic acid, caffeine, respectively. In the above reported research article it was seen that the detection became much easier and less time consuming and need low-cost synthesis way compared to the conventional different spectroscopic method.

Guan et al. reported in his study of used of C-dots prepared from folic acid as precursor in the efficient labeling of C6 cells, indicating the folic acid-based C-dots are a promising nanoprobe for cell imaging.[11] In another study, Lycii Fructus-based C-dots were successfully employed in the bioimaging of Hela cells.[35] Tulsi leaves were also reported for synthesis green fluorescent C-dots and used fluorescence imaging of MDA-MB 468 cells.[18]

Another interesting application is the detection of pesticides with C-dots prepared from different sources. As we know in the conventional farming system, due to the lack of proper knowledge on using of pesticides in terms of way and limit, it is becoming life threatening. Many times it was seen that pesticides directly affect on central nervous system and causes diffident diseases. So use of the pesticides should be within the permissible limit. Also, cross contamination can cause much pollution with pesticides in different water sources starting from drinking water to water used in farm field and effects adversely in the living organism. But detection of the same as discussed earlier needs sophisticated instrument facility. So as there are many reports available on the successful detection of different pesticides with good detection limit with the low-cost fluorescence C-dots, it is becoming the new hope in the research field to develop well-established low-cost method. Burning ash of waste paper was used as a precursor to synthesize C-dots for the detection of chlorpyrifos organophosphorus pesticides.[21] In another reported study commercially available chlorophyll was used to prepare carbon quantum dots for the detection of paraoxon pesticides.[42]

In different fields the C-dots extend its tail and get positive responses. Hence, there is a maximum chance that in few years, the C-dots will be alternate methods for quality and safety evaluation in many areas of food science along with other areas of science.

5.6.1 CARBON DOTS A TOOL FOR DETECTION OF Fe^{3+} ION IN WATER SAMPLES

C-dots showed high sensitivity and selectivity toward metal ions. Based on that the fields of sensor application using C-dots have been significantly expanding. Although C-dots had been successfully and effectively employed for detection of environmentally hazardous metal ions, but it has a peculiar quenching tendency toward the luminescence of C-dots. The possible mechanism of quenching is assumed to be non-radiative electron transfer from the excited state of carbon dots to the d-orbital of Fe^{3+} Along with that, the nitrogen-doped carbon dots play a role in quenching mechanism according to the hypothesis put forward in various studies. The nitrogen-doped carbon dots must be having some coordination interaction with Fe^{3+} because of the occurrence of electron-rich nitrogen atom in the structure. This could result in non-radiative electron transfer between them which can cause a substantial quenching of PL of carbon dots. This electron transfer process plays a major role for the significant quenching of carbon dots fluorescence.[15,23] For detection purpose different concentration from 0 to 100 μL Fe^{3+} solution were added to the aqueous solution of C-dots, and the fluorescence intensities was recorded. It was reported that with the increase of the concentration of Fe^{3+}, the fluorescence intensity of C-dots decreased gradually.[15] Qu et al. reported the detection of Fe^{3+} ion in water sample with C-dots synthesized from dopamine.[32] There are many reports available on the detection of Fe^{3+} with C-dots prepared from Lycii Fructus,[35] bael (*Aegle marmelos*) leaves,[29] papaya,[38] *Citrus sinesis*,[6] etc.

Selectivity of synthesized C-dots of Fe^{3+} ions with other ions:

Selectivity is the major issue to receive a good detection limit. If the other metal ions also interact in the same manner as the metal ion of interest then it will cause problems and results will be in vein. So analyzing selectivity is the major concern for such kind of detection purposes. Hence with different metal ion like Ag^+, Al^{3+}, Ca^{2+}, Cd^{2+}, Cr^{3+}, Co^{2+}, Cu^{2+}, Fe^{2+}, Hg^{2+}, K^+, Mg^{2+}, Mn^{2+}, Na^+, Ni^{2+}, Pb^{2+}, Zn^{2+}, etc are investigated in the same way to search the cation activities in the PL behavior. During the selectivity test for the other metal ion, the Fe^{3+} showed the quenching mechanism whereas other metals did not show any remarkable quenching or no change in fluorescence.[23]

The possible mechanism of that may be the strong bonding and interaction between the Fe^{3+} ions on the surface of C-dots which transferred the photoelectrons from C-dots to Fe^{3+} ions. The coordination interaction of Fe^{3+} and the C-dots was illustrated as the electron in the excited state of C-dots would transfer to the unfilled orbit of Fe^{3+} which leads to nonradioactive electron/hole recombination, resulting in the fluorescence quenching.[15,40]

The study of quenching behavior of C-dots was generally studied by Stern–Volmer equation as stated below (Kalita and Mishra, 2017):

$$F_0/F = 1 + K_{SV}[Q],$$

where F_0 and F represent the fluorescence intensities of C-dots in the absence and presence of metal ions, $[Q]$ is the metal ion concentration and K_{SV} is the Stern–Volmer quenching constant.

For studying the accuracy and validating the proposed method in the study of C-dots, generally samples with C-dots solution are spiked with the metal ion at various concentration and recoveries are calculated by the following formula (Kalita and Mishra, 2017):

$$\text{Recovery\%} = \frac{c_2 - c_1}{c} \times 100$$

where, C is the concentration of Fe^{3+} added into the real samples, C_1 and C_2 is the concentration of Fe^{3+} in the real samples before and after adding the standard Fe^{3+}.

The recoveries of water samples for Fe^{3+} were generally between 96.60 and 108.83%,[15] 100–102% as reported.[29] Hence, the study showed that Citrus limon peel, bael leaves were the excellent sources for synthesizing C-dots and utilized them in the detection of Fe^{3+} ion in water samples. This work provides a new route for low-cost synthesis of bright C-dots from cheap precursors to detect heavy metals like Fe^{3+} with good selectivity and accuracy.

5.6.2   UTILIZATION OF CARBON DOTS FOR VISUAL DETECTION AS A PH PAPER

At different pH range, the fluorescence behavior of C-dots differs. The work had been carried out by Kalita et al. in which they developed C-dots-coated paper for visual fluorescence detection of pH.[15] For that they cut filter papers in a defined size and the pieces were immersed into a C-dots solution for 1 h. After that paper was removed from the solution and placed in an oven at 40°C for 1 h to dry. Different pH solution with pH 4, 5, 6, 7, 8, 9, and 10 were added and dried then observed under UV lamp.[39]

Under UV light, the paper dripped with the acidic solution with pH 2, 4, 6, 9 showed white, slightly darker in color, apparently green, and extra-bright green fluorescence. The raw paper was green under UV light. These studies demonstrated that at acidic range, the carboxyl moieties suspended over the C-dots and quenched the fluorescence but with tending toward the basic pH range quenching effect decreases.[15,39,49] A distinct variation of fluorescence brightness on the paper indicated the rough pH values with naked eyes. Hence can be reported to be utilized the same in preparation of pH paper.

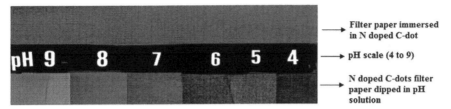

Filter paper immersed
in N doped C-dot

pH scale (4 to 9)

N doped C-dots filter
paper dipped in pH
solution

FIGURE 5.4   Under UV light pH sensitive N doped C-dots immersed filter paper.

5.7   CONCLUSION

It has been shown from the above discussion that the precursor, method of synthesis, mode of synthesis controls the nature of C-dots. They are rich in different properties which can be used in the many field of detection. In the era of popularize use of sophisticated instruments, the sensors develop from this type of low-cost source and method is highly demanding. Also, the other luminescent quantum dots are generally toxic and may be sometimes hazardous to the ecology. But the C-dots, basically prepared from green synthesis are nontoxic with high biocompatibility and non dangerous to the ecology. Hence, the C-dots have reached a remarkable stage only in one decade of discovery and has continuously attracted the interest of researchers from different fields of science and technology. Although there are many challenges, it concurrently opens up new areas of research. The studies to meet the challenges will be beneficial to develop newer scope of C-dots.

KEYWORDS

- carbon dots
- food waste
- fluorescence
- iron
- heavy metal
- pH
- bioimaging
- pesticides
- quantum yield

REFERENCES

1. Ajlec, R.; Štupar, J. Determination of Iron Species in Wine by Ion-Exchange Chromatography-Flame Atomic Absorption Spectrometry. *Analyst* **1989**, *114*, 137−142.
2. Arul, V.; Sethuraman, M. G. Facile Green Synthesis of Fluorescent N-Doped Carbon Dots from Actinidia deliciosa and their Catalytic Activity and Cytotoxicity Applications. *Opt. Mater.* **2018**, *78*, 181–190.
3. Bourlinos, A. B.; Trivizas, G.; Karakassides, M. A.; Baikousi, M.; Kouloumpis, A.; Gournis, D.; Bakandritsos, A.; Hola, K.; Kozak, O.; Zboril, R.; Papagiannouli, I. Green and Simple Route Toward Boron Doped Carbon Dots with Significantly Enhanced Non-Linear Optical Properties. *Carbon* **2015**, *83*, 173–179.
4. Campos, B. B.; Abellán, C.; Zougagh, M.; Jimenez-Jimenez, J.; Rodríguez-Castellón, E.; da Silva, J. E.; Ríos, A.; Algarra, M. Fluorescent Chemosensor for Pyridine Based on N-Doped Carbon Dots. *J. Colloid Interface Sci.* **2015**, *458*, 209-216.
5. Cayuela, A.; Soriano, M. L.; Carrillo-Carrion, C.; Valcarcel, M. Semiconductor and Carbon-Based Fluorescent Nanodots: the Need for Consistency. *Chem. Commun.* **2016**, *52*(7), 1311–1326.
6. Chatzimitakos, T.; Kasouni, A.; Sygellou, L.; Avgeropoulos, A.; Troganis, A.; Stalikas, C. Two of a Kind but Different: Luminescent Carbon Quantum Dots from Citrus Peels for Iron and Tartrazine Sensing and Cell Imaging. *Talanta* **2017**, *175*, 305–312.
7. Chatzimitakos, T.; Kasouni, A.; Sygellou, L;, Leonardos, I.; Troganis, A.; Stalikas, C. Human Fingernails as an Intriguing Precursor for the Synthesis of Nitrogen and Sulfur-Doped Carbon Dots with Strong Fluorescent Properties: Analytical and Bioimaging Applications. *Sens. Actuators B Chem.* **2018**, *267*, 494–501.
8. Dang, D. K.; Sundaram, C.; Ngo, Y. L. T.; Chung, J. S.; Kim, E. J.; Hur, S. H. One Pot Solid-State Synthesis of Highly Fluorescent N and S Co-Doped Carbon Dots and its use as Fluorescent Probe for Ag+ Detection in Aqueous Solution. *Sens. Actuators B Chem.* **2018**, *255*, 3284–3291.
9. De, B.; Karak, N. A Green and Facile Approach for the Synthesis of Water Soluble Fluorescent Carbon Dots from Banana Juice. *RSC Adv.* **2013**, *3*(22), 8286–8290.
10. Deng, J.; Lu, Q.; Mi, N.; Li, H.; Liu, M.; Xu, M.; Tan, L.; Xie, Q.; Zhang, Y.; Yao, S. Electrochemical Synthesis of Carbon Nanodots Directly from Alcohols. *Chem. Eur. J.* **2014**, *20*(17), 4993–4999.
11. Guan, W.; Gu, W.; Ye, L.; Guo, C.; Su, S.; Xu, P.; Xue, M. Microwave-Assisted Polyol Synthesis of Carbon Nitride Dots from Folic Acid for Cell Imaging. *Int. J. Nanomed.* **2014**, *9*, 5071.
12. Guo, Y.; Cao, F.; Li, Y. Solid Phase Synthesis of Nitrogen and Phosphor Co-Doped Carbon Quantum Dots for Sensing Fe^{3+} and the Enhanced Photocatalytic Degradation of Dyes. *Sens. Actuators B Chem.* **2018**, *255*, 1105–1111.
13. He, L.; Liu, C.; Xin, J. H. A Novel Turn-On Colorimetric and Fluorescent Sensor for Fe^{3+} and Al^{3+} with Solvent-Dependent Binding Properties and its Sequential Response to Carbonate. *Sens. Actuators* **2015**, *213*, 181−187.
14. Jelinek, R. *Carbon Quantum Dots.* Springer International Publishing: Cham, 2017; pp 29–46.
15. Kalita, M.; Mishra, P. Synthesis of Carbon Dots from *Citrus limon* Peel by Microwave Assisted Process and its Application for Detection of Ferric Ion (Fe^{3+}) and Development

of pH Paper (Master Degree Major Project), 2017, Department of Food Engineering and Technology, Tezpur University, Tezpur, Sonitpur, Assam, India.

16. Kamal, A.; Kumar, S.; Kumar, V.; Mahajan, R. K. Selective Sensing Ability of Ferrocene Appended Quinoline-Triazole Derivative Toward Fe(III) Ions. *Sens. Actuators* **2015,** *221*, 370–378.

17. Kubojima, Y.; Suzuki, Y.; Tonosaki, M. Vibrational Properties of Japanese Cedar Juvenile Wood at High Temperature. *Bioresour.* **2013,** *8*(4), 5349–5357.

18. Kumar, A.; Chowdhuri, A. R.; Laha, D.; Mahto, T. K.; Karmakar, P.; Sahu, S. K. Green Synthesis of Carbon Dots from Ocimum Sanctum for Effective Fluorescent Sensing of Pb^{2+} Ions and Live Cell Imaging. *Sens. Actuators B Chem.* **2017,** *242*, 679–686.

19. Lee, A. J.; Wang, X.; Carlson, L. J.; Smyder, J. A.; Loesch, B.; Tu, X.; Zheng, M.; Krauss, T. D. Bright Fluorescence from Individual Single-Walled Carbon Nanotubes. *Nano Lett.* **2011,** *11*(4), 1636–1640.

20. Li, H.; Kang, Z.; Liu, Y.; Lee, S. T. Carbon Nanodots: Synthesis, Properties and Applications. *J. Mater. Chem.* **2012,** *22*(46), 24230–24253.

21. Lin, B.; Yan, Y.; Guo, M.; Cao, Y.; Yu, Y.; Zhang, T.; Huang, Y.; Wu, D. Modification-Free Carbon Dots as Turn-on Fluorescence Probe for Detection of Organophosphorus Pesticides. *Food Chem.* **2018,** *245*, 1176–1182.

22. Lin, Y.; Zhou, B.; Martin, R. B.; Henbest, K. B.; Harruff, B. A.; Riggs, J. E.; Guo, Z. X.; Allard, L. F.; Sun, Y. P. Visible Luminescence of Carbon Nanotubes and Dependence on Functionalization. *J. Physical Chem. B* **2005,** *109*(31), 14779–14782.

23. Liu, S.; Liu, R.; Xing, X.; Yang, C.; Xu, Y.; Wu, D. Highly Photoluminescent Nitrogen-Rich Carbon Dots from Melamine and Citric Acid for the Selective Detection of Iron (III) Ion. *Rsc Adv.* **2016,** *6(38),* 31884–31888.

24. Lu, W.; Qin, X.; Liu, S.; Chang, G.; Zhang, Y.; Luo, Y.; Asiri, A. M.; Al-Youbi' A. O.; Sun, X. Economical, Green Synthesis of Fluorescent Carbon Nanoparticles and their Use as Probes for Sensitive and Selective Detection of Mercury (II) Ions. *Anal. Chem.* **2012,** *84*(12), 5351–5357.

25. Lu, W.; Qin, X.; Liu, S.; Chang, G.; Zhang, Y.; Luo, Y.; Asiri, A. M.; Al-Youbi, A. O.; Sun, X. Economical, Green Synthesis of Fluorescent Carbon Nanoparticles and their Use as Probes for Sensitive and Selective Detection of Mercury (II) Ions. *Anal. Chem.* **2012,** *84*(12), 5351–5357.

26. Papanikolaou, G.; Pantopoulos, K. Iron Metabolism and Toxicity. Toxicol. *Appl. Pharmacol.* **2005,** *202*, 199–211.

27. Peng, Z.; Han, X.; Li, S.; Al-Youbi, A. O.; Bashammakh, A. S.; El-Shahawi, M. S.; Leblanc, R. M. Carbon Dots: Biomacromolecule Interaction, Bioimaging and Nanomedicine. *Coord. Chem. Rev.* **2017,** *343*, 256–277.

28. Pooja, D.; Saini, S.; Thakur, A.; Kumar, B.; Tyagi, S.; Nayak, M. K. A "Turn-On" Thiol Functionalized Fluorescent Carbon Quantum Dot Based Chemosensory System for Arsenite Detection. *J. Hazard. Mater.* **2017,** *328*, 117–126.

29. Pramanik, A.; Biswas, S.; Kumbhakar, P. Solvatochromism in Highly Luminescent Environmental Friendly Carbon Quantum Dots for Sensing Applications: Conversion of Bio-Waste into Bio-Asset. *Spectrochimica Acta Part A Mol. Biomol. Spectrosc.* **2018,** *191*, 498–512.

30. Prasannan, A.; Imae, T. One-Pot Synthesis of Fluorescent Carbon Dots from Orange Waste Peels. *Ind. Eng. Chem. Res.* **2013,** *52*(44), 15673–15678.

31. Qin, X.; Lu, W.; Asiri, A. M.; Al-Youbi, A. O.; Sun, X. Microwave-Assisted Rapid Green Synthesis of Photoluminescent Carbon Nanodots from Flour and their Applications for Sensitive and Selective Detection Of Mercury (II) Ions. *Sensors Actuators B Chem.* **2013,** *184*, 156–162.

32. Qu, K.; Wang, J.; Ren, J.; Qu, X. Carbon Dots Prepared by Hydrothermal Treatment of Dopamine as an Effective Fluorescent Sensing Platform for the Label-Free Detection of Iron (III) Ions and Dopamine. *Chem. Eur. J.* **2013,** *19*(22), 7243–7249.

33. Sahu, S.; Behera, B.; Maiti, T. K.; Mohapatra, S. Simple One-Step Synthesis of Highly Luminescent Carbon Dots from Orange Juice: Application as Excellent Bio-Imaging Agents. *Chem. Commun.* **2012,** *48*(70), 8835–8837.

34. Stensballe, A.; Andersen, S.; Jensen, O. N. Characterization of Phosphoproteins from Electrophoretic Gels by Nanoscale Fe(III).Affinity Chromatography with Off-Line Mass Spectrometry. *Anal. Proteom.* **2001,** *1*, 207−222.

35. Sun, X.; He, J.; Yang, S.; Zheng, M.; Wang, Y.; Ma, S.; Zheng, H. Green Synthesis of Carbon Dots Originated from Lycii Fructus for Effective Fluorescent Sensing of Ferric Ion and Multicolor Cell Imaging. *J. Photochem. Photobiol. B Biol.* **2017,** *175*, 219–225.

36. Sun, Y. P.; Zhou, B.; Lin, Y.; Wang, W.; Fernando, K. S.; Pathak, P.; Meziani, M. J.; Harruff, B. A.; Wang, X.; Wang, H.; Luo, P. G. Quantum-Sized Carbon Dots for Bright and Colorful Photoluminescence. *J. Am. Chem. Soc.* **2006,** *128*(24), 7756–7757.

37. Wang, L.; Zhou, H. S. Green Synthesis of Luminescent Nitrogen-Doped Carbon Dots from Milk and its Imaging Application. *Anal. Chem.* **2014,** *86*(18), 8902–8905.

38. Wang, N.; Wang, Y.; Guo, T.; Yang, T.; Chen, M.; Wang, J. Green Preparation of Carbon Dots with Papaya as Carbon Source for Effective Fluorescent Sensing of Iron (III) and Escherichia coli. *Biosens. Bioelectron.* **2016,** *85*, 68–75.

39. Wang, R.; Wang, X.; Sun, Y. One-Step Synthesis of Self-Doped Carbon Dots with Highly Photoluminescence as Multifunctional Biosensors for Detection of Iron Ions and pH. *Sens. Actuators B Chem.* **2017,** *241*, 73–79.

40. Wang, Y.; Dong, L.; Xiong, R.; Hu, A. Practical Access to Bandgap-Like N-Doped Carbon Dots with Dual Emission Unzipped from PAN@ PMMA Core–Shell Nanoparticles. *J. Mater. Chem. C* **2013,** *1*(46), 7731–7735.

41. Wang, Y.; Hu, A. Carbon Quantum Dots: Synthesis, Properties and Applications. *J. Mater. Chem. C* **2014,** *2*(34), 6921–6939.

42. Wu, X.; Song, Y.; Yan, X.; Zhu, C.; Ma, Y.; Du, D.; Lin, Y. Carbon Quantum Dots as Fluorescence Resonance Energy Transfer Sensors for Organophosphate Pesticides Determination. *Biosens. Bioelectron.* **2017,** *94*, 292–297.

43. Xu, Q.; Pu, P.; Zhao, J.; Dong, C.; Gao, C.; Chen, Y.; Chen, J.; Liu, Y.; Zhou, H. Preparation of Highly Photoluminescent Sulfur-Doped Carbon Dots for Fe (III) Detection. *J. Mater. Chem. A* **2015,** *3(2)*, 542–546.

44. Xu, X.; Ray, R.; Gu, Y.; Ploehn, H. J.; Gearheart, L.; Raker, K.; Scrivens; W. A. Electrophoretic Analysis and Purification of Fluorescent Single-Walled Carbon Nanotube Fragments. *J. Am. Chem. Soc.* **2004,** *126*(40), 12736–12737.

45. Yang, Z.; Li, Z.; Xu, M.; Ma, Y.; Zhang, J.; Su, Y.; Gao, F.; Wei, H.; Zhang, L. Controllable Synthesis of Fluorescent Carbon Dots and their Detection Application as Nanoprobes. *Nano-Micro Lett.* **2013,** *5*(4), 247–259.

46. Yang, Z. C.; Wang, M.; Yong, A. M.; Wong, S. Y.; Zhang, X. H.; Tan, H.; Chang, A. Y.; Li, X.; Wang, J. Intrinsically Fluorescent Carbon Dots with Tunable Emission Derived

from Hydrothermal Treatment of Glucose in the Presence of Monopotassium Phosphate. *Chem. Commun.* **2011,** *47*(42), 11615–11617.

47. Zhang, J.; Wen, L.; Miao, F.; Tian, D.; Zhu, X.; Li, H. Synthesis of a Pyridyl-Appended Calix[4]arene and Its Application to the Modification of Silver Nanoparticles as Fe^{3+} Colorimetric Sensor. *New J. Chem.* **2012,** *36,* 656−661.

48. Zhao, S.; Lan, M.; Zhu, X.; Xue, H.; Ng, T. W.; Meng, X.; Lee, C. S.; Wang, P.; Zhang, W. Green Synthesis of Bifunctional Fluorescent Carbon Dots from Garlic for Cellular Imaging and Free Radical Scavenging. *ACS Appl. Mater. Interfaces* **2015,** *7(31),* 17054–17060.

49. Zhong, D.; Miao, H.; Yang, K.; Yang, X. Carbon Dots Originated from Carnation for Fluorescent and Colorimetric pH Sensing. *Mater. Lett.* **2016,** *166*, 89–92.

50. Zhou, L.; Geng, J.; Liu, B. Graphene Quantum Dots from Polycyclic Aromatic Hydrocarbon for Bioimaging and Sensing of Fe^{3+} and Hydrogen Peroxide. *Part. Part. Syst. Charact.* **2013,** *30*, 1086−1092.

51. Zhu, C.; Zhai, J.; Dong, S. Bifunctional Fluorescent Carbon Nanodots: Green Synthesis via Soy Milk and Application as Metal-Free Electrocatalysts for Oxygen Reduction. *Chem. Commun.* **2012,** *48*(75), 9367–9369.

52. Zhu, S.; Song, Y.; Zhao, X.; Shao, J.; Zhang, J.; Yang, B. The Photoluminescence Mechanism in Carbon Dots (Graphene Quantum Dots, Carbon Nanodots, and Polymer Dots): Current State and Future Perspective. *Nano Res.* **2015,** *8*(2), 355–381.

53. Zuo, P.; Lu, X.; Sun, Z.; Guo, Y.; He, H. A Review on Syntheses, Properties, Characterization and Bioanalytical Applications of Fluorescent Carbon Dots. *Microchimica Acta* **2016,** *183*(2), 519–542.

CHAPTER 6

Influence of Drying and Extraction Technology on the Chemical Profile and Antioxidant Property of Mexican Mango Byproduct

ELISEO CÁRDENAS-HERNÁNDEZ, CRISTIAN TORRES-LEÓN*, JUAN ASCACIO-VALDÉS, JUAN C. CONTRERAS-ESQUIVEL, and CRISTÓBAL N. AGUILAR

Research Group on Bioprocesses and Bioproducts, Food Research Department, School of Chemistry, Universidad Autónoma de Coahuila, 25280 Saltillo, Coahuila, México

Corresponding author. E-mail: ctorresleon@uadec.edu.mx

ABSTRACT

The objective of this study was to investigate the obtaining of antioxidant bioactive extracts from the Ataulfo mango seed. The influence of two drying methods (convection and lyophilization) and four extraction techniques (water bath, ultrasound, microwave, and enzymatic) were investigated for their total polyphenol content and antioxidant activity (DPPH and ABTS). The HPLC–MS analysis was also developed to identify the compounds. The results showed that lyophilization drying and ultrasound extraction are the most efficient methods with a phenolic content of 29.45 mg GAE/g and a potent antioxidant activity of 97.47% (DPPH) and 289.87 mg TE/g (ABTS) that was higher than that reported in fruits that are a good source of antioxidants. Enzyme-assisted extraction (EAE) revealed a high polyphenol content and ABTS antioxidant activity, although DPPH antioxidant activity was lower than ultrasound and microwave. This method is proposed as an interesting field of research. HPLC–MS analysis revealed the presence of 16 compounds belonging to the families of lignans, gallotannins, phenolic acids, and flavonols. Our results demonstrate that with the right technology bioactive extracts with a powerful antioxidant activity from by-products of mango processing can be obtained. The extracts have multiple applications in the food and pharmaceutical industry.

6.1 INTRODUCTION

In recent years, the search for bioactive extracts has become a growing trend in scientific research. The bioactive components are food elements that influence cell activity and physiological mechanisms, producing beneficial effects on health, these are generally found in small quantities in products of plant origin and in foods rich in lipids.[1] Fruits such as mango have important phytochemicals, both simple and complex, the most notable are the phenolics acids, mangiferin, carotenoids, and gallotannins.[2]

In general, mango fruit is an important source of nutritional and functional components such as soluble fiber (pectins), vitamins (A, C, and E), bioactive compounds (polyphenols, carotenes), minerals (iron, phosphorus, and calcium), organic acids (citric and malic), among others. However, the peels and seeds of the fruit contain greater amounts of these components, mainly of bioactive components such as polyphenols.[3] Polyphenolic compounds are natural bioactive that have proven biological activities, such as antioxidant, antimicrobial, and anti-inflammatory activity.[4] Scientific evidence also reports an important action in the prevention and treatment of degenerative diseases such as cancer.[5]

In Mexico, mango processing waste constitutes approximately 600,000 tons per year.[6] A medium-sized plant that processes 200 tons of mango per day generates approximately 84 tons of by-products,[7] mainly represented in peels (13–16%)[8] and seeds (10–25%).[3] These by-products are generally discarded in dumps and end up being a source of environmental pollution.

In the world, mango seed has aroused great interest, this by-product is one of the biomaterials with the greatest biological activity.[3] In mango seed extracts from countries such as India, China, Pakistan, and Spain has reported a high content of bioactive compounds,[9] a high antioxidant activity,[10] anticancer activity,[11] antimicrobial activity against Gram-positive and Gram-negative bacteria[12] and antidiarrheal effect.[13] Activities attributed to its high polyphenol content that is higher than that of the pulp and the skin.[14] However, factors such as variety, environmental, and agronomic conditions may affect these properties, so the evaluation of Ataulfo mango seed is a perspective line of research.

Currently, sustainable agricultural production alternatives that allow the use of agro-industrial waste are being sought. A methodology widely used for the valorization of agro-industrial by-products involves the drying (to secure physicochemical and microbiological stability), extraction (since the compounds are bound to constituents of the cell wall), and finally measurement of biological activities (as an activity antioxidant). Two of the used drying methods worldwide are conventional drying in the oven and lyophilization or freeze drying (FD). FD is the most effective way in protecting nutrients during processing, but it is costly. Relatively cheaper conventional drying is commonly used in drying process, but usually results in inferior product quality.[15] Currently, two of the main green technologies investigated in the world for the extraction of compounds are microwave and ultrasonic-assisted extraction: microwave-assisted extraction (MAE) consists in the use of microwaves that generate heat within the material, which leads to faster heating rates and shorter periods of time.[16] In ultrasound-assisted extraction (UAE) the acoustic cavitation causes the cell walls to rupture, intensifying mass transfer and improving the effect of solvent penetration in the plant and capillary tissue.[17] Enzyme-assisted extraction (EAE) is an unconventional method for the extraction of polyphenolic compounds. Pectinex® Ultra is a pectolytic enzyme (pectinase) produced by a strain of *Aspergillus aculeatus* that mainly contains polygalacturonase, pectinase, and pectinesterase, being able to break plant pectic substances.[18] The main purpose of the use of enzymes is to alter the structure of the cell wall and improve the extraction performance, and as a consequence, release phenolic components.[19]

Therefore, identifying the effect of drying methods and extraction technologies to obtain extracts with high biological power is of great importance. The objective of this research work was to determine the effect of drying and extraction techniques in obtaining bioactive extracts with antioxidant activity from the byproduct of Mexican mango seed.

6.2   MATERIALS AND METHODS

6.2.1   CHEMICALS

DPPH (1,10-diphenyl-2-20-picrylhydrazyl), ABTS (2,20-azinobis (3-ethyl-benzothiazoline-6-sulfonic acid)), Trolox (6-hydroxy-2,5,7,8-tetramethyl acid Chroman-2-carboxylic), Folin–Ciocalteu reagent (sodium 3,4-dioxo-3,4-dihydronaphthalene-1-sulfonate), gallic acid (3,4,5-trihydroxybenzoic acid) were purchased from Sigma Chemical Co. (St. Louis, MO, USA), the solvents for HPLC analyzes were of HPLC quality. All other chemicals were used for analytical quality purposes.

6.2.2   BIOLOGICAL MATERIAL AND DRYING METHOD

Ataulfo mango seeds were obtained from the local industry in Saltillo, Coahuila, Mexico. The mango seed was cut into pieces and subsequently dried. At this point two types of drying were evaluated: Convection drying (CD) at 60°C (Felisa, Mexico), the seed was taken to an oven at a temperature of 60°C for 48 h. FD (LABCONCO Freezone 4.5, USA). Lyophilization was performed for 24 h at a temperature of -40°C and a pressure of 50 mPa. The dried mango seeds were sieved to a particle size of 150 μm.

6.2.3   EXTRACTION TECHNIQUES

Extractions of bioactive compounds were carried out by four different methods at a constant temperature and time: Conventional (Water bath), UAE (Branson, USA), MAE (Mars 6, USA), and EAE (Pectinex® Ultra).

The working conditions used for the methods of extraction by water bath, UAE and MAE were the same, only varying the type of drying. Briefly, 1 g of mango seed was weighed and diluted in 30 mL of ethanol (90%), the extraction was performed for 20 min. EAE was developed according to Fernández, Vega, and Aspé.[20] Briefly, 100 mg of seed sample (for each drying method) was diluted in 10 mL of acetate buffer solution at a pH of 3.6 and a concentration of 0.1 M. Subsequently, the enzyme (1% E/S) was added. To finish the experiment, 20 mL of 90% ethanol were added.

The samples were transferred to conical tubes and taken to a centrifuge (SORVALL: Biofuge primo ™ R) at 3000 rpm at a temperature of 20°C and a time of 10 min. Subsequently, the filtering of each sample was carried. The extracts were stored in a refrigerator at a temperature of -20°C.

6.2.4   TOTAL POLYPHENOLS CONTENT

Total phenols were determined spectrophotometrically using the Folin–Ciocalteu reagent.[21] The results were expressed as milligrams of gallic acid equivalents (GAE) per gram of dry plant material using a calibration curve (R^2

= 0.999). 20 µL of the sample was placed in a microplate well. Subsequently, 20 µL of the Folin–Ciocalteu reagent was added and mixed, allowing them to react for 5 min. After this time, 20 µL of sodium carbonate (0.01 M) was added and mixed with a new reaction period of 5 min. Finally, the solution was diluted with 125 µL of distilled water and its absorbance was read at 790 nm in a microplate reader (Epoch, BioTek Instruments, Inc .; Winooski, VT, USA).

6.2.5   DETERMINATION OF ANTIOXIDANT ACTIVITY

6.2.5.1   DPPH RADICAL SCAVENGING ACTIVITY

The DPPH method was performed according to the methodology described by Molyneux [22]. The radical was prepared at a concentration of 60 µM. Subsequently, the absorbance wavelength of the radical was determined by a sweep in the visible range, the methanol-only reading blank was used and as a control absorbance that of the solution DPPH-methanol, 193 µL of DPPH methanolic solution, and 7 µL of sample was added to the microplate. After 30 min of reaction under dark conditions, absorbance was measured at 517 nm, using a spectrophotometer microplate reader (Epoch, BioTek Instruments, Inc.; Winooski, VT, USA) controlled with the Gen5 Data Analysis software interface. The reduction of the DPPH radical was calculated as a percentage of inhibition by eq 6.1.

$$DPPH\ inhibition\,(\%) = \frac{(Ac - As)}{Ac}\,(100) \tag{6.1}$$

where A_c is the control absorbance (DPPH-methanol) and A_s is the absorbance of the sample.

6.2.5.2   ABTS ASSAY

The ABTS method was developed according to Re et al.[23] 1 mL of the ethanol solution of ABTS was mixed with 10 µL of mango seed extract in cuvettes for spectrophotometric analysis. Absorbance was measured at 734 nm in a spectrophotometer (Genesys 20, USA). The results were expressed as mg of Trolox equivalents per gram of dry plant material (mg TE/g). Trolox was diluted to appropriate concentration for the establishment of standard calibration curve (R^2 = 0.998).

6.2.6 IDENTIFICATION OF COMPOUNDS BY HPLC–MS

The samples were taken to mass chromatography to determine the molecular weights and identify the compounds present in the extracts. For this purpose, an HPLC (Varian Pro Star 410, USA) equipped with a pump (Varian Pro Star 230I, USA) and a PDA detector (Varian Pro Star 330, USA) were used. The column used was C18: 150 mm × 2.1 mm, 3μm (Grace, USA). Mass analysis was developed using a mass spectrometer (Varian 500-MS IT Mass Spectrometer, USA).

6.2.7 STATISTICAL ANALYSIS

The experimental design used in the present study corresponded to a completely randomized design with three repetitions for each treatment. The results are presented as mean ± standard deviation. The results were subjected to an analysis of variance and comparison of means using the Tukey test, through Statistica 7 software, with a level of significance p <0.05.

6.3 RESULTS AND DISCUSSION

6.3.1 TOTAL POLYPHENOLS CONTENT

Figure 6.1 shows the results of total polyphenols qualified by the Folin–Ciocalteu method. The drying methods did not show a significant difference in any of the extraction methods so that any of the two drying methods can be used. The use of the most economical method is favored. The extraction by UAE and MAE was statistically better than the water bath. Therefore, either of the two techniques can be used to obtain polyphenolic compounds. The results of phenolic compounds of the present study (29 GAE mg/g) were higher than previously reported in mango seeds from Egypt by Abdel et al.[24] (21.97 mg GAE/g) and Ashoush and Gadallah[25] (23.90 mg GAE/g). Our results confirm that MAE and UAE have better results than traditional methods. These two methods showed statistically better results in the extraction of phenolic compounds from mango seed.

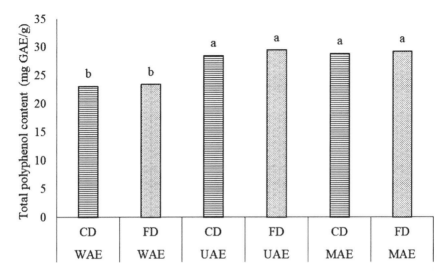

FIGURE 6.1 Effect of different drying and extraction techniques on total polyphenol content in mango seed extracts. (CD: Convection drying; FD: Freeze drying; WAE: Water bath-assisted extraction; UAE: Ultrasound-assisted extraction; MAE: Microwave-assisted extraction).

6.3.2 DPPH RADICAL SCAVENGING ACTIVITY

The antioxidant activity is a parameter that indicates how much the antioxidant compound prevents oxidation of a substrate.[26] The radical DPPH has a purple color when it finds an H^+ with which to complement its structure it loses the color. The change in color is what makes it possible to quantify the antioxidant power of the substance placed as a sample.[22]

As shown in Figure 6.2, the FD drying method has the highest antioxidant activity values compared to CD. The highest values of antioxidant activity were obtained in the combination of FD with MAE and UAE, with values of 97.57% and 97.13%, respectively. In these combinations, a significant difference between the methods was not found. WAE with CD shows the lowest percentage inhibition values (92.81%), a greater DPPH radical inhibitory activity is related to the phenolic content in the extract (Fig. 6.1).

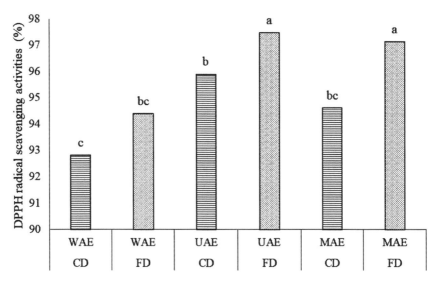

FIGURE 6.2   Effect of different drying and extraction techniques on DPPH radical scavenging property of mango seed extracts (CD: Convection drying; FD: Freeze drying; WAE: Water bath-assisted extraction; UAE: Ultrasound-assisted extraction; MAE: Microwave-assisted extraction).

The results were higher than those determined in mango seed (*Mangifera indica*) from India (79.6%)[27] and mango peel from Peru (76.8%).[28] This percentage of inhibition observed is much higher than the activity reported for other fruits such as cucumber (91.53%),[29] grape (45.0%),[30] apple (61.7 %),[31] and Lima (81.5%).[26] Fruits that are considered a good source of antioxidants.

6.3.3   ABTS ASSAY

Complementary to the DPPH method, the ABTS method is one of the most used methods in the world to identify the antioxidant activity of an extract. ABTS radical has a blue coloration, when it finds an H^+ with which to complement its structure it loses the coloration.[23]

Figure 6.3 shows the antioxidant activity ABTS in equivalent Trolox (ET). The extracts obtained with drying by FD and extraction by UAE and MAE showed the highest values of antioxidant activity. This result is consistent with that determined with the DPPH method. The antioxidant activity of the FD and UAE convictions (289.87 mg TE/g) was significantly higher than that determined for FD and MAE (240.41 mg TE/g). The lowest ABTS antioxidant activity values were obtained by the combination of CD and extraction by water bath (113.13 mg ET/g). The low values determined in the extraction with a water bath is consistent with that determined in the DPPH test and of total phenols. The differences found between the ABTS and DPHH methods may be due to the nature of the reactions, ABTS shows several absorption maxima and good solubility, allowing the test of compounds of both lipophilic and hydrophilic nature, in comparison to DPPH, which can only be dissolved in Organic media.[32]

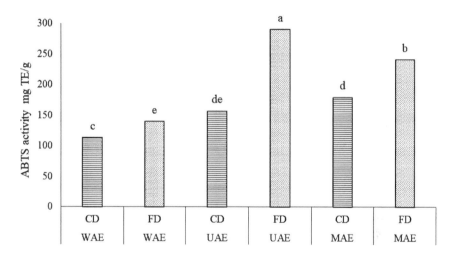

FIGURE 6.3   Effect of drying and extraction techniques on ABTS$^+$ activity of mango seed extracts (CD: Convection drying; FD: Freeze drying; WAE: Water bath-assisted extraction; UAE: Ultrasound-assisted extraction; MAE: Microwave-assisted extraction).

It is noteworthy that the measurement of phenols is not affected by the drying method. However, the antioxidant activity was affected by the drying method. This may be due to the loss of the biological activity of the phenols when dried at a high temperature. The antioxidant activity of DPPH and ABTS shows the influence of drying, obtaining better results with FD. This phenomenon can also be explained by the fact that drying by lyophilization releases the compounds phytochemicals of the cellular structures where they are attached, making them more accessible. In this way, lyophilization not only preserves antioxidant compounds, but also increases their availability.[33] Recent studies have established that the stabilization of the plant material by

FD favors its conservation and also improves the antioxidant activity of the extracts.[10]

6.3.4   POLYPHENOL CONTENT AND ANTIOXIDANT ACTIVITY BY ENZYMATIC EXTRACTION

In Figure 6.4, the results of the polyphenol content and antioxidant activity by ABTS and DPPH for EAE are showed. The analysis of the total phenols shows that the drying method does not show a significant influence on the extracts. This is according to other methods studied. In contrast, the determination of antioxidant activity by ABTS does appear to be influenced by the treatment of drying by FD, since this shows greater antioxidant activity. As we mentioned above, FD can influence the preservation and release of bioactive compounds. In the case of antioxidant activity by the DPPH method, the inhibition rates do not show a significant difference. The polyphenol content and antioxidant activity by ABTS was higher than in the previous treatments, 1019.37 mg GAE/g, and 3686.67 mg TE/g, respectively. However, the DPPH activity was lower (93.45%), this result can be explained by the effect of the enzyme on the extraction. The main purpose of using enzymes is to alter the structure of the cell wall and improve extraction performance, and as a consequence try to release phenolic components more easily.[19] Therefore, the effect of the enzyme is the main reason why high amounts of total polyphenols and antioxidant activity by ABTS are observed compared to the other methods mentioned. It should be noted that this is the first study where EAE is used for the extraction of antioxidants from mango seed, so there is little information, promising results recommend continuing with future work. Additionally, purification treatments are recommended to ensure that other molecules that are released into enzyme treatment do not interfere with spectrophotometric analyses.[34]

6.3.5   IDENTIFICATION BY HPLC–MS

Table 6.1 shows the HPLC–MS identification of phenolic compounds present in mango seed extracts and the effect of the drying method and extraction technique on the compound profile. In total, 16 compounds belonging to the families of lignans, gallotannins, phenolic acids, flavonols, flavonoids, and anthocyanins were identified.

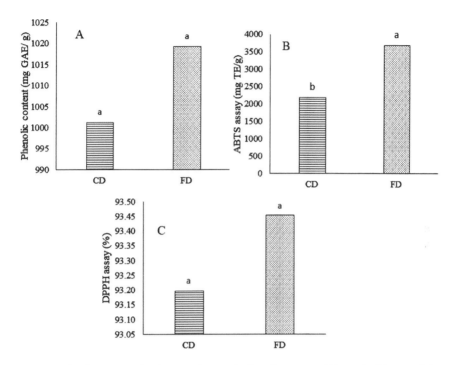

FIGURE 6.4 Enzyme-assisted extraction: polyphenol content (A), antioxidant activity ABTS (B), and antioxidant activity DPPH (C).

In general, the FD method presented the largest number of compounds. This may be because of the preservation of thermo-sensitive components.[35] FD is the best method for obtaining a high-quality dried product.[36] Biological materials are best preserved by FD (lyophilization), a two-step process in which the sample is first frozen and then dried at low temperature under vacuum.[37] In FD, the various heat-sensitive biological compounds are not damaged.[38] The key benefits of FD include the following: retention of biochemical properties and high activity levels.[39] The evaluation of the different combinations of drying methods and extraction techniques showed that UAE and FD has a greater number of bioactive compounds present in the extracts. This relates to the phenols values and antioxidant activity determined. High antioxidant activity can be recognized in the presence of these compounds.

The compounds identified in the extracts have important functional activities, for example, Penta-galloyl glucose (PGG) is a central compound

TABLE 6.1 Identification by HPLC–MS of Phenolic Compounds Present in Mango Seed Extracts According to the Drying Method and Extraction Technique.

Num	RT	Mass	Name	Family	Formula	MAE		UAE		EAE	
						FD	CD	FD	CD	FD	CD
1	2.93	387.1	Medioresinol	Lignans	$C_{21}H_{24}O_7$	•		•		•	
2	3.29	331.2	Galloyl glucose	Gallotannins	$C_{13}H_{16}O_{10}$	•	•	•	•	•	•
3	8.17	169	Gallic acid	Phenolic acids	$C_7H_6O_5$		•		•		•
4	20.54	321	Gallic acid 3-0-gallate	Phenolic acids	$C_{14}H_{10}O_9$		•		•		
5	28.43	787.2	Tetra-galloyl glucose	Gallotannins	$C_{34}H_{28}O_{22}$	•		•		•	
6	28.52	314.9	Ramnetin	Flavonol	$C_{16}H_{12}O_7$			•		•	•
7	28.56	549.4	Quercetin 3-O-(6''-malonila-glucoside)	Flavonol	$C_{24}H_{22}O_{15}$			•		•	•
8	29.25	301.0	Ellagic acid	Phenolic acids	$C_{14}H_6O_8$	•	•	•			
9	30.82	939.2	Penta-galloyl glucose	Gallotannins	$C_{41}H_{32}O_{26}$	•	•	•	•	•	•
10	31.58	545.1	Rhamnetin-3-[6-2-butenoil-hexoside]	Flavonoid	$C_{26}H_{16}O_{13}$						
11	32.38	1091.1	Hexa-galloyl glucose	Gallotannins	$C_{48}H_{36}O_{30}$	•	•	•	•	•	•
12	32.602	775.6	Unidentified					•			
13	33.78	621.1	Apigenin 7-O-diglucoronide	Flavonoid	$C_{27}H_{26}O_{18}$	•	•	•	•	•	•
14	33.78	1243.1	Hepta-galloyl glucose	Gallotannins	$C_{55}H_{40}O_{34}$	•	•	•		•	
15	35.14	697.1	Cyanidine 3-O-(6''-malonil-3''-glucosyl-glucoside)	Anthocyanins	$C_{30}H_{33}O_{19}$	•		•	•	•	
16	36.06	349.0	Ethyl 2, 4-dihydroxy-3-(3, 4, 5-rihydroxybenzoyl) benzoate	Gallotannins	$C_{16}H_{14}O_9$	•		•			
					Total	10	8	13	7	10	7

RT, retention time; FD, freeze drying; CD, convection drying; MAE: microwave-assisted extraction; UAE: ultrasound-assisted extraction; EAE: enzyme-assisted extraction.

in the biosynthetic path of gallotannins that was present in all the evaluated extracts, has been reported as a potent antioxidant.[40] Cell culture-based studies showed that PGG has anticancer activities such as proapoptosis, antiproliferation, antiangiogenesis, metastasis.[41] This compound also has an antiviral effect[42] and antimicrobial effect.[43] PGG has also reported antitumor effects both *in vitro* and *in vivo* in prostate cancer, lung cancer, sarcoma, breast cancer, leukemia, melanoma, and liver cancer.[41] Ellagic acid has a potent antimicrobial activity. Antioxidant, antiviral, and anti-mutagenic properties have also been reported.[44]

The compounds identified in this chapter demonstrate that extracts are a good source of bioactive compounds with high biological activity that may have applications in the industry, for example, ellagic acid is used in the beverage industry (clarification of beverages). Another important application is developed in the wastewater treatment industry where it is used for the precipitation of metals and the reduction thereof.[44] Gallic acid is used in the pharmaceutical industry as an important intermediate compound in the synthesis of the antibiotic trimethoprim; in the chemical industry is used as a substrate for the chemical or enzymatic synthesis of propyl gallate and other antioxidant compounds used in food, cosmetics, hair products, adhesives, and lubricants.[45]

6.4 CONCLUSIONS

Extracts with high antioxidant activity can be obtained from the Ataulfo mango seed. This study showed that the drying method does not significantly influence the total phenolic content. Although the biological activity was affected, the extraction technique that showed a greater obtaining of phenolic compounds and a greater antioxidant activity was the drying by FD and extraction by UAE. The HPLC analysis identified compounds with biological activity from different families such as gallotannins, lignans, flavonoids, and anthocyanins. These compounds can be used industrially in the manufacture of bioactive products. EAE analysis requires further research on the effect of the enzyme on extracts. In general, due to the high antioxidant activity, the extracts have a wide potential to obtain added value and contribute to reducing environmental pollution.

KEYWORDS

- **phenolic compounds**
- **antioxidant activity**
- **lyophilization**
- **mango seed**
- **extraction**
- **microwave**
- **ultrasound**

REFERENCES

1. Fuentes-Barría, H.; Muñoz Peña, D. Influence of the Bioactives Compounds of Beetroot (Beta Vulgaris L) on the Cardioprotective Effect: A Narrative Review. *Rev. Chil. Nutr.* **2018,** *45*(2), 178–182. https://doi.org/10.4067/s0717-75182018000300178.
2. Burton-Freeman, B. M.; Sandhu, A. K.; Edirisinghe, I. Mangos and Their Bioactive Components: Adding Variety to the Fruit Plate for Health. *Food Funct.* **2017,** *8*(9), 3010–3032. https://doi.org/10.1039/c7fo00190h.
3. Torres-León, C.; Rojas, R.; Contreras-Esquivel, J. C.; Serna-Cock, L.; Belmares-Cerda, R. E.; Aguilar, C. N. Mango Seed: Functional and Nutritional Properties. In *Trends in Food Science and Technology*; Elsevier Ltd, 2016; pp 109–117. https://doi.org/10.1016/j.tifs.2016.06.009.
4. Balasundram, N.; Sundram, K.; Samman, S. Phenolic Compounds in Plants and Agri-Industrial by-Products: Antioxidant Activity, Occurrence, and Potential Uses. *Food Chem.* **2006,** *99*(1), 191–203. https://doi.org/10.1016/j.foodchem.2005.07.042.
5. Del rio, D.; Rodriguez-mateos, A.; Spencer, J.; Tognolini, M.; Borges, G.; Crozier, A. Dietary (Poly)Phenolics in Human Health: Structures, Bioavailability, and Evidence of Protective Effects Against Chronic Diseases. *Antioxid. Redox Signal.* **2013,** *18*(14). https://doi.org/10.1089/ars.2012.4581.
6. FAOSTAT. Food and Agriculture Organization of the United Nations. http://www.fao.org/faostat/es/#data.
7. Cock, L. S.; León, C. T. Potencial Agroindustrial de Cáscaras de Mango (Mangifera Indica) Variedades Keitt y Tommy Atkins Agro Industrial Potential of Peels of Mango (Mangifera Indica) Keitt and Tommy Atkins. *Acta Agron.* **2015,** *64*(2), 110–115. https://doi.org/10.15446/acag.v64n2.43579.
8. Serna, L.; Torres, C.; Ayala, A. Evaluation of Food Powders Obtained from Peels of Mango (Mangifera Indica) as Sources of Functional Ingredients. *Inf. Tecnol.* **2015,** *26*(2), 41–50. https://doi.org/10.4067/S0718-07642015000200006.
9. Asif, A.; Farooq, U.; Akram, K.; Hayat, Z.; Shafi, A.; Sarfraz, F.; Sidhu, M. A. I.; Rehman, H. U.; Aftab, S.; Ur-Rehman, H.; et al. Therapeutic Potentials of Bioactive Compounds from Mango Fruit Wastes. *Trends Food Sci. Technol.* **2016,** *53*, 102–112. https://doi.org/10.1016/j.tifs.2016.05.004.

10. Dorta, E.; González, M.; Lobo, M. G.; Sánchez-Moreno, C.; de Ancos, B. Screening of Phenolic Compounds in By-Product Extracts from Mangoes (Mangifera Indica L.) by HPLC-ESI-QTOF-MS and Multivariate Analysis for Use as a Food Ingredient. *Food Res. Int.* **2014,** *57,* 51–60. https://doi.org/10.1016/j.foodres.2014.01.012.

11. Noratto, G. D.; Bertoldi, M. C.; Krenek, K.; Talcott, S. T.; Stringheta, P. C.; Mertens-Talcott, S. U. Anticarcinogenic Effects of Polyphenolics from Mango (Mangifera Indica) Varieties. *J. Agric. Food Chem.* **2010,** *58,* 4104–4112. https://doi.org/10.1021/jf903161g.

12. Khammuang, S.; Sarnthima, R. Antioxidant and Antibacterial Activities of Selected Varieties of Thai Mango Seed Extract. *Pak. J. Pharm. Sci.* **2011,** *24*(1), 37–42. https://doi.org/10.1016/j.ifset. 2009.10.004.

13. Rajan, S.; Suganya, H.; Thirunalasundari, T.; Jeeva, S. Antidiarrhoeal Efficacy of Mangifera Indica Seed Kernel on Swiss Albino Mice. *Asian Pac. J. Trop. Med.* **2012,** *5*(8), 630–633. https://doi.org/10.1016/S1995-7645(12)60129-1.

14. Ajila, C.; Naidu, K.; Bhat, S.; Rao, U. Bioactive Compounds and Antioxidant Potential of Mango Peel Extract. *Food Chem.* **2007,** *105*(3), 982–988. https://doi.org/10.1016/j.foodchem.2007.04.052.

15. Que, F.; Mao, L.; Fang, X.; Wu, T. Comparison of Hot Air-Drying and Freeze-Drying on the Physicochemical Properties and Antioxidant Activities of Pumpkin (Cucurbita Moschata Duch.) Flours. *Int. J. Food Sci. Technol.* **2008,** *43*(7), 1195–1201. https://doi.org/10.1111/j.1365-2621.2007.01590.x.

16. Martínez-Ramírez, A.; Carlos Contreras-Esquivel, J.; Belares-Cerda, R. Extracción de Polifenoles Asistida Por Microondas a Partir de Punica Granatum L. *Acta Quim. Mex.* **2010,** *2*(4), 1–5.

17. Guntero V., Longo M. V., Ciparicci S., Francescato F., SanMartino M., Martini R., A. A. *Comparación de Métodos de Extracción de Polifenoles a Partir de Residuos de La Industria Vitivinícola.*; San Francisco, Cordoba, 2015.

18. Sánchez-Madrigal, M. Á.; Viesca-Nevárez, S. L.; Quintero-Ramos, A.; Amaya-Guerra, C. A.; Meléndez-Pizarro, C. O.; Contreras-Esquivel, J. C.; Talamás-Abbud, R. Optimization of the Enzyme-Assisted Extraction of Fructans from the Wild Sotol Plant (Dasylirion Wheeleri). *Food Biosci.* **2018,** *22,* 59–68. https://doi.org/10.1016/j.fbio.2018.01.008.

19. Flores, E. Extracción de Antioxidantes de Las Bayas Del Sauco (Sambucus Nigra L. Subsp. Peruviana) Con Ultrasonido, Microondas, Enzimas y Maceración Para La Obtención de Zumos Funcionales. *Inf. Tecnol.* **2017.** https://doi.org/10.4067/S0718-07642017000100012.

20. Fernández, K.; Vega, M.; Aspé, E. An Enzymatic Extraction of Proanthocyanidins from País Grape Seeds and Skins. *Food Chem.* **2015,** *168,* 7–13. https://doi.org/10.1016/j.foodchem.2014.07.021.

21. Wong, J.; Muñiz, D.; Aguilar, P.; Rodríguez, R.; Aguilar, C. Microplate Quantification of Total Phenolic Content from Plant Extracts Obtained by Conventional and Ultrasound Methods. *Phytochem. Anal.* **2014,** *25*(5), 439–444. https://doi.org/10.1002/pca.2512.

22. Molyneux, P. The Use of the Stable Free Radical Diphenylpicryl-Hydrazyl (DPPH) for Estimating Antioxidant Activity. *Songklanakarin J. Sci. Technol.* **2004,** *26*(2), 211–219. https://doi.org/10.1016/S0891-5849(98)00315-3.

23. Re, R.; Pellegrini, N.; Proteggente, A.; Pannala, A.; Yang, M.; Rice-Evans, C. Antioxidant Activity Applying an Improved ABTS Radical Cation Decolorization Assay. *Free Radic. Biol. Med.* **1999**. https://doi.org/10.1016/S0891-5849(98)00315-3.

24. Abdel, M.; Ashoush, I.; Nessrien, M. Characteristics of Mango Seed Kernel Butter and its Effects on Quality Attributes of Muf Fi Ns. *Food Sci. Technol.* **2012**, *9*(2), 1–9.

25. Ashoush, I.; Gadallah, M. Utilization of Mango Peels and Seed Kernels Powders as Sources of Phytochemicals in Biscuit. *World J. Dairy Food Sci.* **2011**, *6*(1), 35–42. https://doi.org/20113260380.

26. Pérez-Nájera, V.; Lugo-Cervantes, E.; Gutiérrez-Lomelí, M.; Del-Toro-Sánchez, C. L. Extracción de Compuestos Fenolicos de La Cáscara de Lima (Citrus Limetta Risso) y Determinación de Su Actividad Antioxidante. *Biotecnia* **2013**, *15*(3), 18. https://doi.org/10.18633/bt.v15i3.153.

27. Hussah Abdullah, A. S.; Sabo Mohammed, A.; Abdullah, R.; Elwathiq Saeed Mirghani, M. Identification and Quantification of Phenolic Compounds in Mangifera Indica Waterlily Kernel and Their Free Radical Scavenging Activity. *J. Adv. Agric. Technol.* **2015**, *2*(1), 1–7. https://doi.org/10.12720/joaat.2.1.1-7.

28. Jibaja Espinoza Luis Miguel. Determinación de La Capacidad Antioxidante y Análisis Composicional de La Harina de Cáscara de Mango (Mangífera Indica) Variedad "Criollo" Procedente de La Provincia de Sullana En Piura. *Tecnol. Desarro.* **2015**, *13*(1), 23–26. https://doi.org/https://doi.org/10.18050/td.v13i1.748.

29. Sudha, G.; Priya, M. S.; Shree, R. I.; Vadivukkarasi, S. In Vitro Free Radical Scavenging Activity of Raw Pepio Fruit (Solanum Muricatum Aiton). *Int. J. Curr. Pharm. Res.* **2011**, *3*(2), 137–140.

30. Jayaprakasha, G. K.; Girennavar, B.; Patil, B. S. Radical Scavenging Activities of Rio Red Grapefruits and Sour Orange Fruit Extracts in Different in Vitro Model Systems. *Bioresour. Technol.* **2008**, *99*(10), 4484–4494. https://doi.org/10.1016/j.biortech.2007.07.067.

31. Leja, M.; Mareczek, A.; Ben, J. Antioxidant Properties of Two Apple Cultivars during Long-Term Storage. *Food Chem.* **2003**, *80*(3), 303–307. https://doi.org/10.1016/S0308-8146(02)00263-7.

32. Kuskoski, E. M.; Asuero, A. G.; Troncoso, A. M.; Mancini-Filho, J.; Fett, R. Aplicación de Diversos Métodos Químicos Para Determinar Actividad Antioxidante En Pulpa de Frutos. *Ciência e Tecnol. Aliment.* **2005**, *25*(4), 726–732. https://doi.org/10.1590/s0101-20612005000400016.

33. Dorta, E.; Lobo, M. G.; González, M. Using Drying Treatments to Stabilise Mango Peel and Seed: Effect on Antioxidant Activity. *LWT - Food Sci. Technol.* **2012**, *45*(2), 261–268. https://doi.org/10.1016/j.lwt.2011.08.016.

34. Muñoz-bernal, O.; Torres-aguirre, G. A. Nuevo Acercamiento a La Interacción Del Reactivo de Folin-Ciocalteu Con Azúcares Durante La Cuantificación de Polifenoles Totales. *TIP Rev. Espec. en Ciencias Químico-Biológicas* **2017**, *20*(2), 23–28. https://doi.org/10.1016/j.recqb.2017.04.003.

35. Punathil, L.; Basak, T. Microwave Processing of Frozen and Packaged Food Materials: Experimental. *Ref. Modul. Food Sci.* **2016**, 1–28. https://doi.org/10.1016/B978-0-08-100596-5.21009-3.

36. Marques, L. G.; Silveira, A. M.; Freire, J. T. Freeze-Drying Characteristics of Tropical Fruits. *Dry. Technol.* **2006**, *24*(4), 457–463. https://doi.org/10.1080/07373930600611919.

37. Bhambere, D.; Gaidhani, K.; Harwalkar, M.; Nirgude, P. Freeze Drying- A Review. *World J. Pharm. Res.* **2015,** *4*(8), 516–543.

38. Krokida, M. K.; Philippopoulos, C. Volatility of Apples during Air and Freeze Drying. *J. Food Eng.* **2006,** *73*(2), 135–141. https://doi.org/10.1016/j.jfoodeng.2005.01.012.

39. Ciurzyńska, A.; Lenart, A. Freeze-Drying - Application in Food Processing and Biotechnology - a Review. *Polish J. Food Nutr. Sci.* **2011,** *61*(3), 165–171. https://doi.org/10.2478/v10222-011-0017-5.

40. Kiss, A.; Naruszewicz, M. Polyphenolic Compounds Characterization and Reactive Nitrogen Species Scavenging Capacity of Oenothera Paradoxa Defatted Seed Extracts. *Food Chem.* **2012,** *131*(2), 485–492. https://doi.org/10.1016/j.foodchem.2011.09.011.

41. Shaikh, Q. U. A.; Yang, M.; Memon, K. H.; Lateef, M.; Na, D.; Wan, S.; Eric, D.; Zhang, L.; Jiang, T. 1,2,3,4,6-Pentakis[-O-(3,4,5-Trihydroxybenzoyl)]-α,β-D-Glucopyranose (PGG) Analogs: Design, Synthesis, Anti-Tumor and Anti-Oxidant Activities. *Carbohydr. Res.* **2016,** *430*, 72–81. https://doi.org/10.1016/j.carres.2016.04.021.

42. Haid, S.; Grethe, C.; Bankwitz, D.; Grunwald, T.; Pietschmann, T. Identification of a Human Respiratory Syncytial Virus (HRSV) Cell Entry Inhibitor by Using a Novel Lentiviral Pseudotype (HRSVpp) System. *J. Virol.* **2015,** *90*(6), 3065–3073. https://doi.org/10.1128/JVI.03074-15.

43. Jiamboonsri, P.; Pithayanukul, P.; Bavovada, R.; Chomnawang, M. T. The Inhibitory Potential of Thai Mango Seed Kernel Extract against Methicillin-Resistant Staphylococcus Aureus. *Molecules* **2011,** *16*(8), 6255–6270. https://doi.org/10.3390/molecules16086255.

44. Francisco Virgilio, C. A.; Saúl, S. P.; Martinez-Vázquez, G., Aguilera, A.; Cristóbal Noé Aguilar, R. R. y. Propiedades Químicas e Industriales Del Ácido Elágico. *Acta Quim. Mex.* **2010,** *3*(2), 12.

45. Rodríguez-Durán Luís, V.; Valdivia-Urdiales, B.; Contreras-Esquivel, J. C.; Cristóbal N. Aguilar, R. R. H.y. Química y Biotecnología de La Tanasa. *Acta Quim. Mex.* **2010,** *2*(4), 10.

CHAPTER 7

Food Loss and Waste in the Circular Bioeconomy

S. M. GARCÍA-SOLARES[1,2], C. A. GUTIÉRREZ[3], E. E. NERI-TORRES[3], and I. R. QUEVEDO[3*]

[1]*Centro Mexicano para la Producción más Limpia, Instituto Politécnico Nacional, Av. Acueducto s/n, Col. La Laguna Ticomán, Ciudad de México 07340, México*

[2]*Laboratorio Nacional de Desarrollo y Aseguramiento de la Calidad de Biocombustibles (LaNDACBio), C.P. 07340, Ciudad de México, México*

[3]*Departamento de Ingeniería Química Industrial y de Alimentos (DIQIA), Universidad Iberoamericana Ciudad de México (UIA). Prolongación Paseo de la Reforma 880, Santa Fe, Col. Contadero, C.P. 01219, Ciudad de México, México*

[*]Corresponding author. E-mail: ivan.quevedo@ibero.mx

ABSTRACT

The term food losses and waste (FLW) distinguishes between food losses occurring in the supply chain versus food waste; generated in the distribution and final consumption. The composition of the FLW comprises carbohydrates, lipids, and proteins, representing an economic alternative to be used as raw material in different bioprocesses that might generate value-added products. Different authors have proposed that the key solution for the food industry resides in finding the appropriate approaches to valorize the FLWs. Obtaining high value-added products from FLWs aligns with the current concept of sustainable development aiming to achieve food security, environmental protection, and energy efficiency. This chapter makes a compilation of main studies to quantify and valorize FLW within a circular bioeconomy.

7.1   INTRODUCTION

The term food losses and waste (FLW) is ambiguous and it depends on the context where the terminology is applied (i.e., Food and Agriculture Organization of the United Nations (FAO), European Commission, United States Environmental Protection Agency, or the World Resources Institute).[1] However, within the scope of this manuscript, the definition of FLW distinguishes between food losses (FL) occurring in the supply chain vs. food waste; generated in the distribution and final consumption. In general, FLW relates to the decrease in the quantity or quality of food through the food supply chain (FSC).[2-6]

FLW could be classified as avoidable (edible) or inevitable (inedible) (e.g., bones, seeds, fruit, and vegetable peels) (Fig. 7.1). However, it must be noticed that the characteristics of the waste are influenced by the geography, user culture, and socioeconomic factors. For instance, while in certain communities, some discards of poultry, such as viscera and legs, are part of the FSC, in others will not be the case.[7]

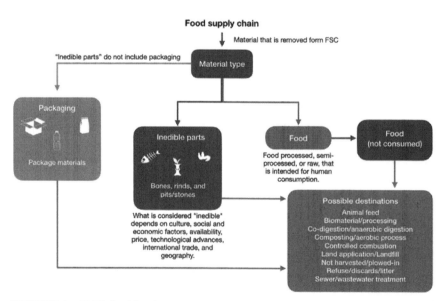

FIGURE 7.1   FLW classification.

The composition of the FLW comprises carbohydrates (glucose), lipids, and proteins, representing an economic alternative to use as raw material in different bioprocesses that might generate value-added products (e.g., biofuels, bioplastics).[8-12] Or even in the implementation of a biorefinery (Section 7.5.4), where the avoidable and unavoidable FLW can be incorporated, potentiating the use and obtaining high-value products which might promote the approach of a circular bioeconomy and generating less damage to the environment in both, short and long term.[9]

FLW is potentialized in crop production, marketing, distribution, and purchases, without influencing the conditions of the country, mainly due to the food policies that each country obeys (e.g., aesthetic standards and expiration dates). That is why for various reasons different solutions must be formulated that meet each of the conditions of the countries interested in reducing and taking advantage of the FLW.[13]

Recently, important efforts have been made to reduce the amount of FLW and to integrate it to eradicate hunger and increase food safety. FLW generates unnecessary expenses and resources such as water (since 70% of freshwater is destined for this activity), energy, working hours, and mineral fertilizers affecting the economy and threatening the livelihoods of future generations.[3,14]

In addition to the economic problems generated by FLW, they also generate greenhouse gases that in developing countries account for 15–28% of the total emissions.[15,16] Different strategies have been implemented to reduce the FLW, in the regulatory framework promoting greater sensitivity for consumers regarding the origin of food, production, man-hours cost, energy, water, and money involved in its production.

The 2030 Agenda for Sustainable Development has the greatest impact worldwide and in its objective "12," indicates "Ensure sustainable consumption and production modalities" and the overall goals of the said objective; "3-Reduce global food waste per capita by half in retail and consumer level by half and reduce food losses in production and distribution chains," "5-Substantially reduce waste generation through prevention policies, reduction, recycling and, reuse," which clearly contributes to the minimization of FLW.

The fulfillment of objective 12 of the 2030 agenda can be facilitated through a circular bioeconomy policy by achieving sustainability in the consumption and production of resources by reducing greenhouse gas emissions.[9] Likewise, it has as priority the efficient conversion of biomass, the use of renewable materials at industrial level and the minimization of food waste, through sustainable agricultural production and food security.[17,18]

The reduction and reuse of the FLW have gained a great interest due to the environmental, economic, and social issues associated. However, generating awareness among consumers is not enough, and a systematic investigation and quantification of each region must be carried out including socioeconomic and environmental impacts of reduction strategies.

7.2　INTERNATIONAL PROTOCOLS FOR THE MANAGEMENT OF FOOD LOSS AND WASTE

FLW has a high economic impact estimated at $940 billion per year and exacerbates food insecurity. Several protocols have been proposed to quantify FLW. In these studies, different factors that cause food losses (FL) and food waste (FW, see Section 7.1) have been identified. These factors are contained in the FSC described below in detail.

7.2.1　FSC

The FSC is divided into the following stages: (1) preharvest food production, (2) postharvest food production, (3) processing, (4) distribution, (5) retail, (6) services, and (7) consumption.[19] Therefore, several protocols have been developed for the study of FLW situations. In this case, the studies take a different criterion (i.e., economic level of the country and production of raw materials) for the quantification. These criteria are acquaintance as the actors that have influence in the FSC.

7.2.2　ACTORS IN THE FSC

It is important to make a correct mapping of the actors when this process is carried out. The United States is one of the world powers in food science and technology which has developed a complete map. The study by the European Union[20] says that these actors in the FSC reach an exorbitant amount (up to 11 million farmers and 300 thousand enterprises) consequently; there are losses and food wastes. According to FAO[7] (see Section 7.1), food loss incorporates the products that leave a production chain to be integrated into another or as an outflow, and food waste which are counted as the foods that have been processed, without quality, contaminated with an external agent, or expired shelf life.

Some of the important actors are producers and farmers, food processors, distribution channels (brokers and retail). These participants can increase the price of a raw material even in triplicate, which is a limitation for the correct development of the FSC.

7.2.3   QUANTIFICATION OF FOOD LOSS AND WASTE

There are different methodologies to quantify FLW, which have as main objectives: (1) to save economic resources to farmers and food-producing companies, (2) to make FLWs smaller for the opportunity to feed more people, and (3) to reduce FLW resulting in less use of water, land, and climate change will decrease. These objectives align with the Sustainable Development Goals and in particular, the objective "12" Responsible consumption and production.

7.2.4   PROTOCOL FOOD LOSS AND FOOD WASTE ACCOUNTING AND REPORTING (STANDARD VERSION 1.0)

This protocol has the objective of quantifying food and nonedible parts, which is known as FLW, with this protocol it is sought an international quantification standard. Since important criteria are given to carry out this process to be considered in each region of the planet.[20]

7.2.5   PROTOCOL CHAMPIONS 12.3

It should be mentioned that this coalition has developed FLW protocols for situations of quantification in restaurants,[21] catering services,[22] and hotels[23]. It is very important to recognize that each of the FSC actors have different points of importance. However, the main factor of analysis is cost–benefit studies. Financial gains (low costs and additional profits) include project management which determines the appropriate quantities of raw material (economic order quantity) on which the inventories, storage, and final production (year, month, day) that allow a study of times and movements for the decrease of FLW.

7.2.6   FOOD LOSS AND FOOD WASTE REDUCTION PROTOCOL BY WRAP ORGANIZATION

The WRAP organization has developed an FLW reduction protocol at households. It has been determined that some of the factors that generate the increase in FLW are as follows: how is food sold and the price structure? Shelf life, how is food stored in the home? And the ways in which food is prepared and served at home. This organization has developed a simulation model through discrete events to determine how the characteristics of a product are related to the decisions of the person who stores it in the home that is life and shelf with the locations that the product takes at home (packaging, product opening, preservation by cooling, waste) the method can be used to explore the interaction between the different environments to which a product is exposed through weather. The structure of the model is given by the following modules: purchase, storage, demand, consumption. This is a basic prediction model through probabilistic events; however, it has the limitation that it can only make predictions on the amount of FLW, it cannot make precise estimates for other variables.

On the other hand, a report of FLW in Mexico is presented. The objectives are to understand the challenges associated with FLW in Mexico, find the main sources of FLW and the reasons that cause it, and finally identify and map potential solutions that can be implemented in the short, medium, and long term.

In this report FLW has been estimated for 20 million tons, the associated economic cost is US $25 billion (representing about 2.5% of Mexico's GDP), however, the data can be much higher, so it is recommended to make other estimates more accurately.

7.2.7   WORLD BANK MEXICO

Mexico is one of the countries with more food production with 286 million tons, but it also has several hotspots due to the amount of FLW that has been quantified. Therefore, within the established protocols, the World Bank has developed a study with the objective of determining the FLW in the state of Jalisco. In this study, the amount of FLW for 79 important products has been determined.[24] In Table 7.1, we can see the results.

As it is possible to observe in Table 7.1, the average percentage of food waste is at least 38%, which translates into approximately 12 million tons of waste of the main foods, which is alarming as in Mexico, there is no entrenched culture of recycling or reuse. However, the study carried out proposes methods of FLW reduction and waste recovery.[24]

TABLE 7.1   Level of FLW in Mexico.

Food	Weekly consumption (t or m³)	Annual consumption (t/m³)	Adding consumption in restaurants, hotels, schools, hospitals	Annual per capita consumption	Waste (t)	Waste (%)
Tortilla	160,587	8,373,470	9,956,056	79.9	2,857,388	28.7
Beef	23,191	1,209,272	1,559,961	12.5	552,382	35.41
Pork	1543	80,459	102,988	0.8	41,391	40.19
Ham	2636	137,423	147,043	1.2	57,200	38.90
Chicken	15,716	819,489	926,023	7.4	275,955	29.80
Fish	4589	239,294	256,044	2.1	99,115	38.71
Shrimp	451	23,512	31,271	0.3	15,257	48.79
Milk	164,586	8,584,800	10,645,152	85.4	4,590,189	43.12
Potatoes	37,244	1,943,789	2,079,854	16.7	788,057	37.89
Avocado	13,327	695,539	785,959	6.3	312,812	39.80
Tomato	34,232	1,784,964	2,356,153	18.9	925,968	39.30
Nopal	3374	175,908	205,813	1.7	76,768	37.30
Mango	20,126	1,050,378	1,176,424	9.4	468,570	39.83
Apples and pears	9176	478,988	560,416	4.5	218,170	38.93
Guava	2962	154,449	168,349	1.4	63,687	37.83
Papaya	6967	363,270	430,475	3.5	171,458	39.83

Given the above quantification protocols, it is worth mentioning that many of them only focus on quantification. However, there are several efforts to add value to the FLW, an issue that will be worked on in the next part of the chapter.

7.3   METHODOLOGIES TO QUANTIFY FOOD LOSS AND FOOD WASTE

To achieve the implementation and monitoring of the decrease in FLW as well as their use in a society aimed at a circular bioeconomy, it is necessary to have a, solid scientific knowledge and technologies of sustainability circles, allowing a balance between society and the ecosystems. Science and technology have provided numerous answers to these challenges.

The need to account for FLW arises from the fact that FAO in 2009 estimated that one-third by weight of all food produced worldwide is wasted. To estimate the impact that the FLW have, various methodologies have been designed for quantification and the footprints that generate at the economic and environmental level.

An accurate quantification of FLW is complex as it must take into account the following variables: (1) stages of the supply chain to the consumer by geographic area; (2) food policies and socioeconomic extracts; (3) designing a mechanism for each variable that allows obtaining real data, with the intention of giving a specific and sustainable solution to each of the problems presented during the aforementioned variables; and (4) an inventory of the FLW of the region and included in the circular bioeconomy.

The development of effective methodologies for the quantification of FLW must include: (1) the general and specific objectives, (2) numerical methods to measure and separate, limits of the system that allow to obtain representative data, international standardized definitions of the terminology of the FLW consistently applied, time periods (season, year, sampling time), destination at the end of the FLW, statistically establish the uncertainties of accounting, the acquisition of primary data (e.g., the composition and generation at retail levels), and secondary data (e.g., literature in which a case study for the quantification of FLW is reported).[25]

There are several reports with FLW quantification methodologies that use a case study for a specific geological area and stage of the FSC, commonly using primary and secondary data from FAO statistical yearbooks.[2,6,7,25–31]

The main disadvantages of these reports are biased results, data cannot be collected throughout the FSC of all generators and are limited to a geological area with specific socio-economic customs.

However, the World Resources Institute has promoted the "*Protocol of lost and wasted food*" being the first document considered worldwide standard for measuring food loss and waste, which indicates 10 quantification methods, including weighing, surveys, composition, and mass balance calculation, among others, also report key assumptions for quantification such as the sources for obtaining primary and secondary data and the calculation of uncertainty.[32]

Another study used worldwide for the quantification of FLWs is that of Gustavsson, Cederberg, and Sonesson,[5] which uses primary and secondary data from the FAO 2007 national and regional food balance sheets.

There are several programs worldwide that FAO and other instances have generated for the systematic, standardized, and worldwide accounting of FLW, some of which are listed below.

FAO, International Food Policy Research Institute and the research program in policies, institutions, and markets of the consortium of research centers (CGIAR), made available to multiple actors the Technical Platform for Quantification and Reduction of FLW which aims to exchange information about accounting for FLW.

A study was conducted in the 27 countries of the European Union, which investigated the causes, figures and environmental impacts of FLW.

The African Postharvest Losses Information estimates the FLW of seven cereal crops during the postharvest food supply stage, in handling, conservation, storage, and transportation processes.

In the United States, the Economic Research Service of the Department of Agriculture has established a data system that allows FLW to be estimated based on the total food supply.

In October 2014, 13 Latin American and Caribbean countries formed a network of experts to establish solutions to the FLW, through the dissemination of knowledge and the approach of various methodologies dissemination of knowledge, methodologies, and innovation among countries.

In parallel, the scientific community has contributed with specific case studies that indicate methodologies for accounting, reuse, recycling, and use of FLW.[33-36]

For the minimization and use of FLW, it is necessary to raise awareness among consumers at any stage of the supply chain, especially in developed countries so that they adopt more efficient technologies to avoid or reduce FLW, as well as the development and application of policies that contribute to this problem.

As a first approximation of the solution to the FLW is to carry out a strategy that is carried out from the homes through the planning of food purchases in accordance with the needs of each one.

7.4   ADVANCED VALORIZATION STRATEGIES FOR FOOD LOSS AND FOOD WASTE

An increase in FLW generation is likely a direct consequence of the availability of food to the increasing populations worldwide and to the economic development, and consumerism in the world today.[37] The food industry of the twenty-first century faces important challenges to maintain its competitiveness due to new laws, regulations, and standards imposed by the current approach to propose a sustainable management agenda world-wide. In developing countries, most of the FLW is disposed in landfills which is eventually causing environmental and health concerns.[38] Different authors have proposed that the key solution for the food industry resides in finding appropriate approaches to valorize the FLWs.[15] Obtaining high value-added products from FLWs align with the current concept of sustainable development aiming to achieve food security, environmental protection, and energy efficiency. Thus, combined efforts have been proposed in recent years to exploit food waste as a bioresource for our next generation of energy, chemical, pharmaceutical, cosmetic, food, and other high value-added products.[38]

7.4.1   THERMOCHEMICAL PROCESSING

Thermochemical processing of FLWs uses heat and catalysts to transform biopolymers into chemical, fuels, or electrical power with a significant volume reduction and minimal waste streams.[39] Although thermochemical processing is a mature technology in the commercial exploitation of hydrocarbons, the complicated chemistries of plant molecules have not been fully explored for biobased feedstocks.[39] In general, thermochemical processing occurs rapidly at high temperatures above ambient conditions (up to 1000°C). The process is characterized as voracious in the pace of reaction and the large variety of materials contained in the FLWs (e.g., carbohydrate, but lignin, lipids, proteins, and other plant compounds), yet its selectivity may not be as indiscriminate as initially proposed. For instance, Refs. [40,41] reported that under controlled conditions, the thermal depolymerization of cellulose (i.e., polysaccharide) mainly results on the production of Levoglucosan; and in a second study that lignin depolymerizes to phenolic compounds (i.e., phenol, 4-vinyl phenol, 2-methoxy-4-vinyl phenol, and 2,6-dimethoxy phenol) as the major products. Thus, since most thermochemical processing will occur under high-temperature combustion and gasification it is expected the chemical equilibrium among the products obtained, namely fuels, chemical, or power.[39] Thermochemical processing may split into four different categories depending on the amount of air used during the process and the characteristics of the FLW (Fig. 7.2).

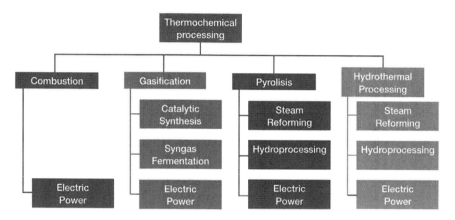

FIGURE 7.2   Thermochemical options for production of fuels, chemicals, and power.[39]

7.4.1.1   COMBUSTION

Combustion of biomass is the rapid reaction of the biomolecule with oxygen to obtain thermal energy and combustion gas, consisting primarily of carbon dioxide and water.[39] Depending on the characteristics of the FLW used, the amount of air used to burn the biomass, and the construction of the furnace, flame temperatures would operate between moderate to high temperatures (800–1600°C) which would be suitable enough for electric power generation.[42] Moreover, as previously mentioned, direct combustion has the advantage that it employs commercially well-developed technology for conventional fossil fuels, as it is the most common approach for electric power generation around the world. Yet, the thermochemical processing of FLW by combustion possesses important disadvantages: as it will result in the formation of greenhouse gasses due to the high operating temperatures, and the difficulties associated to provide enough feedstock to modern electric power plants.[42]

7.4.1.2   GASIFICATION

Gasification is a process of thermal conversion of biomass into a valuable gaseous product called syngas and a solid product, called char.[43] Gasification generates both moderate temperature thermal energy (700–1000°C) and the syngas obtained is a flammable gas mixture, which can be used to generate either electric power or to synthesize fuels or other chemicals using catalysts or even microorganisms (syngas fermentation) (Fig. 7.3).

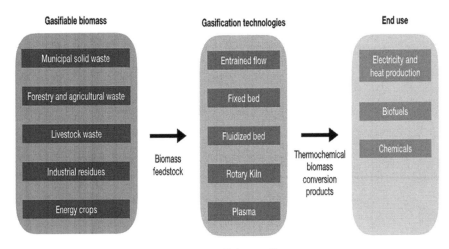

FIGURE 7.3   Thermochemical conversion of biomass.[43]

7.4.1.3   PYROLYSIS

Pyrolysis processes occur at moderate temperatures (300–600°C) in the absence of oxygen and at atmospheric pressure.[39] In the specific case of fast pyrolysis, high heating rates followed by rapid cooling and condensation of the vapors and aerosols result in an energy-rich liquid known as bio-oil.[44] Then the obtained bio-oil might be burned for electric power generation. In comparison to the previously described thermochemical processing, it must be underlined that pyrolysis has the potential to connect the conventional agricultural business to the energy industry, as well as integrating biological processes at the industrial scale.[45] For instance, dedicated biorefineries to convert lignin residues to bio-oil or fermentation of the fast pyrolysis sugar fraction (Fig. 7.4).

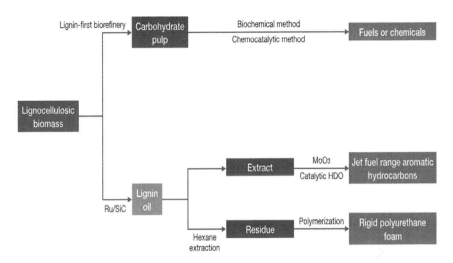

FIGURE 7.4   Thermochemical conversion of biomass.[45]

7.4.1.4   HYDROTHERMAL LIQUEFACTION

Hydrothermal liquefaction is meant to treat wet biomass materials to produce carbohydrates, liquid hydrocarbons, or gaseous products depending upon the reaction conditions without drying and to access ionic reaction conditions by maintaining a liquid water processing medium.[39] Hydrothermal processing is ideal for wet feedstocks (5–20 wt% solids) at temperatures ranging between 200 and 500°C, and high pressures (50–250 atm) to prevent boiling the water contained in the slurry to extract biocrude, syngas, or fractionated plant polymers.[46] However, despite hydrothermal liquefaction has been established as a feasible process for valorizing a variety of agricultural and FLW, its use in efficient industrial applications still represents a challenge before being a commercially viable technology.

7.4.2   BIOCHEMICAL PROCESSING

In recent years, different alternatives have been proposed to study the biochemical conversion of FWL to biofuels, chemicals, biodegradable polymers, and chemical intermediates.[8] As FLW is a rich source of proteins, carbohydrates (hemicellulose, cellulose, starch, and sugar like sucrose, fructose, and glucose), oil, mineral, and fat that can be used in a wide range of enzymatic and microbial processes. Mainly by two options: anaerobic digestion (AD) and fermentation.

7.4.2.1   ANAEROBIC DIGESTION

The high organic matter and the water content of FWLs can be valorized by biological treatments.[47] AD is one technique that may be used to maximize the substrate and bioenergy accumulated in FWLs. In general terms, AD refers to the conversion of organic matter into methane and in the absence of oxygen. The methane (CH_4) produced by the reaction could then be used as fuel or to generate electrical power. Anaerobic digestion consists broadly of three phases: enzymatic hydrolysis, acid formation, and gas production (Fig. 7.5).

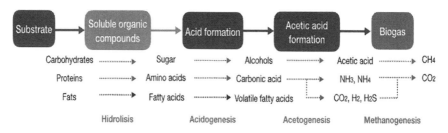

FIGURE 7.5   Anaerobic digestion phases.[47]

The potential of FWL for biomethanation will directly depend on its composition (i.e., dry matter, total ammonia nitrogen, cellulose content, carbon-to-nitrogen (C/N) ratio).[16] Moreover, the digestate coming from AD could be used as fertilizer. In a recent study[48] reported the development and use of a digestate processing system at the pilot scale from the digestate of a mesophilic dairy manure digester and a thermophilic food waste digester for crop production at the farm-scale on a tomato crop.

7.4.2.2   FERMENTATION

Fermentation is a metabolic process in which an organism produces chemical changes in an organic substrate under aerobic or anaerobic conditions. Ethanol is one of the major products by yeast fermentation that has a high demand as a fuel blend. However, recently the waster valorization for the extraction of other products (i.e., lactic acid, succinic acid, butanol, and 3-hydroxybutyrate) has gained considerable interest.[49]

7.4.3   VALUE-ADDED PRODUCTS

Bioplastics are valuable biomaterials that can be produced using fermentation of food waste. Biodegradable and nonbiodegradable biological polymers represent approximately 8.3% of total production[50] that is an estimate of 0.5% of world consumption.[51] The most dynamic development is planned for bioPET, whose production capacity was around 600,000 t in 2013 and it is projected to reach about 7 million tons by 2020, using bioethanol from sugarcane. Bio-based PET production is expanding at high rates worldwide, in large part due to the Plant Technology Collaborative initiative launched by Coca-Cola.[50] The second most dynamic development is planned for polyhydroxyalconates (PHAs), biopolymers that can be obtained from waste and that can be suitable to replace petroleum-derived plastics.[15] Conventional production of PHAs involves high-cost carbon sources such as glucose. Thus, food waste has been recommended as a low-cost carbon source to produce PHAs. Table 7.2 summarizes the properties of FWL applied in PHA production.[52]

TABLE 7.2   Properties of Potential FWLs in PHA Production.

No.	FWL	Origin	Characteristics
1	Used cooking oil	Palm oil, rapeseed oil, soybeans, sunflower seeds	High lipids content that can be converted into fatty acid methyl esters
2	Animal by-products	Blood, fats, residues from intestines	High nitrogen content or high levels of BOD and COD
3	Organic crop residues	Straw, stover, peels, fruit pomace	These fractions consist of important sources of sugars, lipids, carbohydrates, and mineral acids
4	Mixed domestic waste	Cheese whey, waste bread, nuts, and nutshells	High protein content, starch, fat, and fatty acids

7.4.4   INTEGRATED PROCESSING

Integrated processing is a new approach where alternative valorization options could be used for the processing of FWLs, and the by-products obtained from each process could serve as raw material or energy source for the other. Figure 7.6 shows a flowchart of conceptual integrated bioprocessing (biorefinery) for food waste management prepared using the available knowledge from various resources. Since by-products and waste streams from one process remain within the biorefinery boundary and are fed as an input to another process, there will be limited waste flow out of the biorefinery. The appropriate use of this concept is essential to attain the sustainability goals in a circular economy.

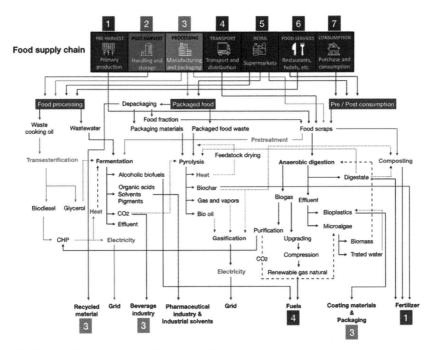

FIGURE 7.6   Conceptual biorefinery for FWL valorization.

7.5   CONCLUSION

FLW is a complex problem that has an impact on global food security and reducing it can generate benefits like save money for all actors involved in the FSC, increasing the quantity of food that is available for human consumption and diminish the environmental impacts.

Several FLW quantification protocols have been developed for different institutions (e.g., WRI, WRAP, IFPRI, USDA, etc.). The development of an effective FLW quantification methodology must include: general and specific project goals, adequate methods to separate, and measure FLWs, set system limits that allow to obtain representative data, use standardized definitions of the FLW international terminology, define time periods for the study, identify the final disposition of FLWs, data statistical analysis to establish the uncertainties and the acquisition of primary and secondary data. Moreover, the quantification study must take into consideration the FSC variables (e.g., geographical region, the stages of FSC involved in the study, etc.).

In recent years, important efforts have been proposed to exploit food waste as a bioresource for our next generation of energy, chemical, pharmaceutical, cosmetic, food, and other high value-added products, which might promote the approach of a circular bioeconomy. Besides, obtaining high value-added products from FLWs aligns with the current concept of sustainable development.

A potential technology for processing FLW is thermochemical processing. This process may split into four different categories depending on the amount of air used and the characteristics of the FLW: combustion, gasification, pyrolysis, and hydrothermal liquefaction. Other alternatives for biochemical conversion of FWL to biofuels, chemicals, biodegradable polymers, and chemical intermediates are anaerobic digestion and fermentation.

In the case of bioplastics, those are valuable biomaterials that can be produced using fermentation of food waste. More specifically, the PHAs can be obtained from waste and that can be suitable to replace petroleum-derived plastics.

The technologies described above could be unified in an integrated processing, where alternative valorization options could be used for the processing of FWLs and the byproducts obtained from each process could serve as raw material o energy source for the other.

KEYWORDS

- **food loss and waste**
- **food supply chain**
- **bioeconomy**
- **bioprocess engineering**
- **waste valorization**
- **waste management**
- **sustainable development**

REFERENCES

1. Thyberg, K. L.; Tonjes, D. J. Drivers of Food Waste and Their Implications for Sustainable Policy Development. *Resour., Conserv. Recycl.* **2016,** *106,* 110–123. doi:10.1016/j.resconrec.2015.11.016.
2. Parfitt, J. , Barthel, M.; MacNaughton, S. Food Waste within Food Supply Chains: Quantification and Potential for Change to 2050. *Philos. Trans. Royal Soc. B: Biol. Sci.* **2010,** *365,* 3065–3081. doi:10.1098/rstb.2010.0126.
3. Kummu, M.; de Moel, H.; Porkka, M.; Siebert, S.; Varis, O.; Ward, P. J. Lost Food, Wasted Resources: Global Food Supply Chain Losses and Their Impacts on Freshwater, Cropland, and Fertiliser Use. *Sci. Total Environ.* **2012,** *438,* 477–489. doi:10.1016/j.scitotenv.2012.08.092.
4. Van Der Werf, P.; Gilliland, J. A. A Systematic Review of Food Losses and Food Waste Generation in Developed Countries. *Proc. Inst. Civil Eng.: Waste Resource Manage.* **2017,** *170,* 66–77. doi:10.1680/jwarm.16.00026.
5. Gustavsson, J.; Cederberg, C.; Sonesson, U. Global Food Losses and Food Waste. *Unep* **2011,** 1. doi:10.1098/rstb.2010.0126.
6. Poças Ribeiro, A.; Rok, J.; Harmsen, R.; Rosales Carreón, J.; Worrell, E. Food Waste in an Alternative Food Network—A Case Study. *Resour., Conserv. Recycl.* **2019,** *149,* 210–219. doi:10.1016/j.resconrec.2019.05.029.
7. Food and Agriculture Organization of the United Nations (FAO). *Food Wastage Footprint, Impacts on Natural Resources,* 2013.
8. Uçkun Kiran, E.; Trzcinski, A. P.; Ng, W. J.; Liu, Y. Bioconversion of Food Waste to Energy: A Review. *Fuel* **2014,** *134,* 389–399. doi:10.1016/j.fuel.2014.05.074.
9. Mak, T. M. W.; Xiong, X.; Tsang, D. C. W.; Yu, I. K. M.; Poon, C. S. Sustainable Food Waste Management Towards Circular Bioeconomy: Policy Review, Limitations and Opportunities. *Bioresour. Technol.* **2019,** 122497. doi:10.1016/j.biortech.2019.122497.
10. Raza, Z. A.; Abid, S.; Banat, I. M. Polyhydroxyalkanoates: Characteristics, Production, Recent Developments and Applications. *Int. Biodeterior. Biodegrad.* **2018,** *126,* 45–56. doi:10.1016/j.ibiod.2017.10.001.
11. Chen, G. Q.; Jiang, X. R. Engineering Bacteria for Enhanced Polyhydroxyalkanoates (PHA) Biosynthesis. *Synth. Syst. Biotechnol.* **2017,** *2,* 192–197. doi:10.1016/j.synbio.2017.09.001.
12. Rivero, C. P.; Hu, Y.; Kwan, T. H.; Webb, C.; Theodoropoulos, C.; Daoud, W.; Lin, C. S. K. Current Developments in Biotechnology and Bioengineering. **2017.** doi:10.1016/B978-0-444-63664-5.00001-0.
13. HLPE. *Las Pérdidas y el Desperdicio de Alimentos en el Contexto de Sistemas Alimentarios Sostenibles.* 2014; p 133. http://www.fao.org/cfs/cfs-hlpe/informes/es/.
14. Lemaire, A.; Limbourg, S. How Can Food Loss and Waste Management Achieve Sustainable Development Goals? *J. Cleaner Prod.* **2019,** *234,* 1221–1234. doi:10.1016/j.jclepro.2019.06.226.
15. Ravindran, R.; Jaiswal, A. K. Exploitation of Food Industry Waste for High-Value Products. *Trends Biotechnol.* **2016,** *34,* 58–69. doi:10.1016/j.tibtech.2015.10.008.
16. Fisgativa, H.; Tremier, A.; Dabert, P. Characterizing the Variability of Food Waste Quality: A Need for Efficient Valorisation Through Anaerobic Digestion. *Waste Manage.* **2016,** *50,* 264–274. doi:10.1016/j.wasman.2016.01.041.

17. Schütte, G. What Kind of Innovation Policy Does the Bioeconomy Need? *New Biotechnol.* **2018**, *40*, 82–86. doi:10.1016/j.nbt.2017.04.003.
18. Venkata Mohan, S.; Chiranjeevi, P.; Dahiya, S. A.; Kumar, N. Waste Derived Bioeconomy in India: A Perspective. *New Biotechnol.* **2018**, *40*, 60–69. doi:10.1016/j.nbt.2017.06.006.
19. CCA. Caracterización y Gestión de la Pérdida y el Desperdicio de Alimentos en América del Norte, 2017. http://webcache.googleusercontent.com/search?q=cache:http://www3.cec.org/islandora/en/item/11772-characterization-and-management-food-loss-and-waste-in-north-america-es.pdf.
20. WRI. Food Loss and Waste Accounting and Reporting Standard. *FLW Protocol* **2016**, *160*.
21. Champions 12.3. *The Business Case for Reducing Food Loss and Waste: Restaurants.* 2019; pp 1–14.
22. Champions 12.3. *The Business Case for Reducing Food Loss and Waste: Catering Service. Report on Behalf of Champions 12.3*, 2019; pp 1–14.
23. Champions 12.3. *The Business Case for Reducing Food Loss and Waste: Hotels*, 2019; pp 1–14.
24. Kemper, K.; Hickey, V.; Ahuja, P. S.; Kneller, C. *Mexico Conceptual Framework on Food Loss and Waste*, n.d.
25. Corrado, S.; Caldeira, C.; Eriksson, M.; Hanssen, O. J.; Hauser, H. E.; van Holsteijn, F.; Liu, G.; Östergren, K.; Parry, A.; Secondi, L.; Stenmarck, Å.; Sala, S. Food Waste Accounting Methodologies: Challenges, Opportunities, and Further Advancements. *Global Food Secur.* **2019**, *20*, 93–100. doi:10.1016/j.gfs.2019.01.002.
26. Xue, L.; Liu, G. *Introduction to Global Food Losses and Food Waste*; Elsevier Inc.: Amsterdam, Netherlands, 2019. doi:10.1016/b978-0-12-815357-4.00001-8.
27. Plazzotta, S.; Manzocco, L. *Food Waste Valorization*; Elsevier Inc.: Amsterdam, Netherlands, 2019. doi:10.1016/b978-0-12-815357-4.00010-9.
28. Teigiserova, D. A.; Hamelin, L.; Thomsen, M. Review of High-Value Food Waste and Food Residues Biorefineries with Focus on Unavoidable Wastes from Processing. *Resour., Conserv. Recycl.* **2019**, *149*, 413–426. doi:10.1016/j.resconrec.2019.05.003.
29. Hartikainen, H.; Mogensen, L.; Svanes, E.; Franke, U. Food Waste Quantification in Primary Production—The Nordic Countries as a Case Study. *Waste Manage.* **2018**, *71*, 502–511. doi:10.1016/j.wasman.2017.10.026.
30. FAO. *Pérdidas y Desperdicio de Alimentos en el Mundo—Alcance, Causas y Prevención*, 2012. doi:10.3738/1982.2278.562.
31. Comisión-para-la-Cooperación-Ambiental. *Por Qué y Cómo Cuantificar la Pérdida y el Desperdicio de Alimentos*, 2019.
32. WRI. Estándar de Contabilización y Reporte Sobre Pérdida y Desperdicio de Alimentos. *Resumen Ejecut* **2016**, *17*. http://flwprotocol.org/wp-content/uploads/2016/05/FLW-Standard-executive-summary-SPANISH.pdf.
33. Salihoglu, G.; Salihoglu, N. K.; Ucaroglu, S.; Banar, M. Food Loss and Waste Management in Turkey. *Bioresour. Technol.* **2018**, *248*, 88–99. doi:10.1016/j.biortech.2017.06.083.
34. Shafiee-Jood, M.; Cai, X. Reducing Food Loss and Waste to Enhance Food Security and Environmental Sustainability. *Environ. Sci. Technol.* **2016**, *50*, 8432–8443. doi:10.1021/acs.est.6b01993.

35. Muth, M. K.; Birney, C.; Cuéllar, A.; Finn, S. M.; Freeman, M.; Galloway, J. N.; Gee, I.; Gephart, J.; Jones, K.; Low, L.; Meyer, E.; Read, Q.; Smith, T.; Weitz, K.; Zoubek, S. A Systems Approach to Assessing Environmental and Economic Effects of Food Loss and Waste Interventions in the United States. *Sci. Total Environ.* **2019,** *685,* 1240–1254. doi:10.1016/j.scitotenv.2019.06.230.

36. Principato, L.; Ruini, L.; Guidi, M.; Secondi, L. Adopting the Circular Economy Approach on Food Loss and Waste: The Case of Italian Pasta Production. *Resour. Conserv. Recycl.* **2019,** *144,* 82–89. doi:10.1016/j.resconrec.2019.01.025.

37. Xue, L.; Liu, G.; Parfitt, J.; Liu, X.; Van Herpen, E.; Stenmarck, Å.; O'Connor, C.; Östergren, K.; Cheng, S. Missing Food, Missing Data? A Critical Review of Global Food Losses and Food Waste Data. *Environ. Sci. Technol.* **2017,** *51,* 6618–6633. doi:10.1021/acs.est.7b00401.

38. Ong, K. L.; Kaur, G.; Pensupa, N.; Uisan, K.; Lin, C. S. K. Trends in Food Waste Valorization for the Production of Chemicals, Materials and Fuels: Case Study South and Southeast Asia. *Bioresour. Technol.* **2018,** *248,* 100–112. doi:10.1016/j.biortech.2017.06.076.

39. Brown, R. C. Introduction to Thermochemical Processing of Biomass into Fuels, Chemicals, and Power. *Thermochemical Processing of Biomass: Conversion into Fuels, Chemicals and Power*, 2011; pp 1–12. doi:10.1002/9781119990840.ch1.

40. Patwardhan, P. R.; Satrio, J. A.; Brown, R. C.; Shanks, B. H. Product Distribution from Fast Pyrolysis of Glucose-Based Carbohydrates. *J. Anal. Appl. Pyrol.* **2009,** *86,* 323–330. doi:10.1016/j.jaap.2009.08.007.

41. Patwardhan, P. R.; Brown, R. C.; Shanks, B. H. Understanding the Fast Pyrolysis of Lignin. *ChemSusChem* **2011,** *4,* 1629–1636. doi:10.1002/cssc.201100133.

42. Jenkins, B. M.; Baxter, L. L.; Koppejan, J. *Biomass Combustion*, 2011. doi:10.1002/9781119990840.ch2.

43. Molino, A. Chianese, S. Musmarra, D. Biomass Gasification Technology: The State of the Art Overview. *J. Energy Chem.* **2016,** *25,* 10–25. doi:10.1016/j.jechem.2015.11.005.

44. Venderbosch, R. H.; Prins, W. *Fast Pyrolysis*; 2011. doi:10.1002/9781119990840.ch5.

45. Huang, Y. Duan, Y. Qiu, S. Wang, M. Ju, C. Cao, H. Fang, Y. Tan, T. Lignin-First Biorefinery: A Reusable Catalyst for Lignin Depolymerization and Application of Lignin Oil to Jet Fuel Aromatics and Polyurethane Feedstock. *Sustain Energy Fuels* **2018,** *2,* 637–647. doi:10.1039/c7se00535k.

46. Elliott, D. C. *Hydrothermal Processing*, 2011. doi:10.1002/9781119990840.ch7.

47. Paritosh, K.; Kushwaha, S. K.; Yadav, M.; Pareek, N.; Chawade, A.; Vivekanand, V. Food Waste to Energy: An Overview of Sustainable Approaches for Food Waste Management and Nutrient Recycling. *BioMed Res. Int.* **2017,** *2017.* doi:10.1155/2017/2370927.

48. Barzee, T. J.; Edalati, A.; El-Mashad, H.; Wang, D.; Scow, K.; Zhang, R. Digestate Biofertilizers Support Similar or Higher Tomato Yields and Quality Than Mineral Fertilizer in a Subsurface Drip Fertigation System. *Front. Sustain. Food Syst.* **2019,** *3,* 1–13. doi:10.3389/fsufs.2019.00058.

49. Waqas, M.; Rehan, M.; Khan, M. D.; Nizami, A.-S. *Conversion of Food Waste to Fermentation Products*; Elsevier: Amsterdam, Netherlands, 2019. doi:10.1016/b978-0-08-100596-5.22294-4.

50. Nova-Institut. Bio-Based Building Blocks and Polymers in the World, Market Study on Bio-Based: Polymers in the World. *Ind. Biotechnol.* **2013,** *11,* 360. doi:10.1089/ind.2015.28999.fae.

51. Iles, A.; Martin, A. N. Expanding Bioplastics Production: Sustainable Business Innovation in the Chemical Industry. *J. Cleaner Prod.* **2013,** *45*, 38–49. doi:10.1016/j.jclepro.2012.05.008.
52. Tsang, Y. F.; Kumar, V.; Samadar, P.; Yang, Y.; Lee, J.; Ok, Y. S.; Song, H.; Kim, K. H.; Kwon, E. E.; Jeon, Y. J. Production of Bioplastic Through Food Waste Valorization. *Environ. Int.* **2019,** *127*, 625–644. doi:10.1016/j.envint.2019.03.076.

CHAPTER 8

Assessment and Analysis of the Fruit and Vegetable Losses Due to Its Transportation in Mexico City Wholesale Market

ELISEO GARCÍA-PÉREZ[1], GENARO AGUILAR-GUTIÉRREZ[2], and ALEJANDRA RAMÍREZ MARTÍNEZ[1*]

[1]*Postgrado en Agroecosistemas Tropicales, Colegio de Postgraduados campus Veracruz, Colegio de Postgraduados Campus Veracruz, Veracruz 91700, Mexico*

[2]*Sección de Estudios de Posgrado e Investigación, Escuela Superior de Economía, Instituto Politécnico Nacional, Miguel Hidalgo, Ciudad de México 11340, Mexico*

[*]*Corresponding author. E-mail: ramirez.alejandra@colpos.mx*

ABSTRACT

There is scarce information on the loss of fruits and vegetables registered in the wholesale market of Mexico City, one of the biggest wholesale markets in the world. Considering the above, it is important to generate data on the losses generated in Mexico City wholesale market and to analyze the economic impact of these losses in the region. To achieve this objective, we visited the fruit and vegetable sector. We obtained the data of the food loss registered by the administrators of the visited stores, which was obtained by the difference in the weight of the truck that transported the product. We also interviewed persons in charge of transport and the administrators/transporters of the visited stores to assess the measures taken during the transportation and selling of the products. The economic impact was quantified by a formulae adapted for the situation. We also carried out an analysis of the transportation in the food loss. Results confirmed that a great quantity of fruit and vegetables are loss and that this loss has a huge impact in the region. To our knowledge, this is the first time such information will be available to the scientific community and the public.

8.1 WHOLESALE MARKETS IN THE WORLD

Some studies have identified parts of the food chain where food losses and waste often occur (Fehr et al., 2002; Parfitt et al., 2010; Koester et al., 2013; Bagherzadeh et al., 2014; Chaboud, 2017). Particularly, wholesale markets are important places to study food loss because they concentrate a great quantity of food products so that there is a risk that a great quantity of food may be lost if it is not properly transported and stocked, and these markets may be a component of the food chain that has been disregarded (Cadhilon et al., 2000).

Since ancient times, different types of markets served to trade foods and other goods in different parts of the world (Braudel, 1979). In Mexico, for example, Spanish chronists described big markets, known as *tianguis*, which were bigger than the cities in their own country and this practice still persists these days (Villamar, 2016). Since ancient times, these markets are important places not only to trade foods and other goods but also to build social interactions between the buyers (Villamar, 2016; Linares and Bye, 2016). The influence of these ancient markets in Mexico is so important that it seems that the marketing in contemporary Mexico is a consequence of the relationships established during different historical periods (Linares and Bye, 2016). In Europe, markets similar to wholesale markets have been in existence since the XVII century. But it was after World War II that the modern wholesale markets were created to supply fruits, vegetables, and other food to the population (Cadhilon et al., 2000). Moreover, until 1991, very few new wholesale markets had been created in developed countries. In the USA, some wholesale markets became "food centers" and now include other nonfresh food products. As in the USA, wholesale markets have suffered from a declining role in the distribution of food into Western European cities (Tracey-White, 1991). However, their basic function of providing food to small retailers, catering firms, and institutional customers in the city centers has not disappeared (Cadhilon et al., 2000).

8.2   FOOD LOSSES STUDIES AT THE TRANSPORTATION AND WHOLESALE MARKET LEVEL

In general, there is scarce data about food losses specifically due to the transportation/distribution at the wholesale level. Since the 1970s, losses of fruit and vegetables were attributed to improper transportation, particularly, the lack of proper refrigeration during the transportation process (Harvey, 1978). Afterward, Gustavsson et al. (2013) conducted a study for the Food and Agriculture Organization (FAO) of the United Nations regarding food losses to include the losses due to the distribution stage. In this case, the distribution steps comprised the transportation from the field to the retailers and from the retailers to final consumers. Losses during distribution varied between 4% and 15% for fruit and vegetables, meat, and fish depending on the region and product group. However, according to Jedermann et al (2014), they did not consider cold chain losses as a separate category. Thus, according to Jedermann et al. (2014), part of the category called "post-harvest handling and storage" with losses of between 0.5% and 10% should be added to the distribution losses. In the same order of ideas, greater transportation losses are observed in developing countries when compared to developed countries. Beretta et al. (2013), for example, reported that the postharvest handling and trade loss (including transportation) of apples was around 1% due to the high technical standards established in Switzerland. In addition, fruit and vegetable losses between a center of a major retailer and a producer were relatively small: between 0.35% and 0.44%.

The actual food loss at the wholesale level in different countries is shown in Table 8.1. Since the establishment of the FAO, the reduction of food loss was within its mandate (Parfitt et al., 2010). A pioneer study, carried out by Kling (1943), revealed that wholesale waste could amount to almost one percent of total food production. He associated that the losses in these markets were due to the incorrect storage of foods, its distribution, and spoilage caused by insufficient movement of goods and improper handling. In more recent years, Kitinoja and Al-Hassan (2012) assessed the loss of fruits and vegetables through the weights of these fruit/vegetable packages. They found that in Africa, the loss could be as high as 17.9% and in India 11%. The main reasons for the losses were high temperatures during the harvest, handling, transport and marketing of the studied crops, poor packaging quality, poor field sanitation, and the time required to reach the retail market.

In the case of Egypt and Colombia, the amount of tomatoes loss greatly differs from each other. Interestingly, the study developed in Egypt revealed, at the same time, great differences in food loss values when it was assessed through interviews from when they were assessed directly (5.2% versus 17.9%, respectively). Chaboud (2017) estimated the averages for transport losses of tomatoes transported without refrigeration. They found that losses were equal to zero due to relatively fast delivery and proper handling. As mentioned above, refrigeration, proper handling, and the time of transportation have been already identified by researchers and FAO (Gustavsson et al., 2013) as the key factors in reducing food loss; however, the study of Chaboud (2017) suggests that refrigeration could not be necessary if foods are transported in short distances.

TABLE 8.1 Summary of the Research Studying the Food Loss at the Distribution and Wholesale Level in Different Countries.

Country	Food product	Food loss value (%)	References
27 EU member countries*	General/Not specified	4.5	Koester (2014)
Ghana	Nine fruit and vegetable crops	12.9	Kitinoja and Al-Hassan (2012), Saran et al. (2012)
Rwanda	Four fruit and vegetable crops	16.2	Kitinoja and Al-Hassan (2012), Saran et al. (2012)
Benin	Seven fruit and vegetable crops	17.9	Kitinoja and Al-Hassan (2012), Saran et al. (2012)
Egypt	Potato, tomato, and grapes	Potatoes: 1.5 Tomatoes: 17.9 Grapes: 6.9	Blond (1984)
India	Six fruit and vegetable crops	11	Kitinoja and Al-Hassan (2012), Saran et al. (2012)
Brazil	Fruit and vegetables	4	Fehr et al. (2002)
Colombia*	Tomatoes	<2 (wholesale level) 0 (transportation)	Chaboud (2017)

*This value represents both the food loss at the wholesale/retail level.

*The specific value of the food loss at the wholesale level is not specified.

Studies have shown that in the case of the European Union members, food loss at the wholesale level may be lower than in other parts of the world because they tend to use appropriate technologies to handle, transport, and store food (Tracey-White, 1991). In fact, in these countries, the largest amount of food loss is due to consumer habits (Prusky, 2011). In the case of Brazil, the reasons for the amount of fruit and vegetable loss at wholesale level were not deeply discussed. The study realized by Fehr et al. (2002) suggests that the problem of the food loss at the wholesale level is not negligible. However, it is outweighed by the loss registered at the retail level.

Now, we will analyze the case of Mexico's wholesale market.

8.3   STUDY'S METHODOLOGY

Food loss due to transportation in the Mexico City's wholesale market was obtained from the records of the differences in the weight of the trucks that transport the fruits and vegetables when they arrived at the wholesale market. The main objective of this research was to obtain direct data of the food loss of fruit and vegetables during the transportation losses. To know the factors that may enhance food loss once food arrived at the wholesale market, our interviews included questions about the daily care of the fruits and vegetables they sold. These interviews were carried out during a week in September 2018. We only visited the fruit and vegetable sector and selected those stores that offered their products in the aisles. We interviewed 10 personnel in charge of the transport of the fruits and vegetables in the Mexico City's Wholesale Market (CACM), and 30 administrators of some stores at the CACM.

The questions that were asked to the sellers/administrators were as follows:

- Is the truck used to transport the goods purchased or rented?
- Why do you lose product?
- How much money do you think is lost?
- What care is given to the product once it arrives at the wholesale market?
- How often did you resupply your store?
- Do you take advantage of the product that is not sold?

The questions that were asked to the responsible for the transportation of food were as follows:

- How long does it take to transport the fruits and vegetables you distribute?
- What care is given to the product during transport?
- What is the origin of your merchandise?

All statistical analyses were performed using the Minitab 16 software. Care data and types of fruit were analyzed as dummy variables.

8.4   MEXICO CITY'S WHOLESALE MARKET

CACM is one of the largest wholesale markets in Latin America and in the world for its extension (327 ha) and the volume of the products sold (972,000 t in 2003) (Robles-Martínez et al., 2010; Morales-Pérez, 2011; INIFAP, 2010; World Union of Wholesale Markets, 2019). The CACM is open 24/7 and comprises eight markets: (1) fruits and vegetables, (2) flowers and vegetables, (3) groceries, (4) jamaiquita, (5) poultry and meat, (6) auction and producers, (7) empty containers, and (8) fish. As a terminal wholesale market, the CACM specializes in big sales (mostly by the ton), but also dispenses to smaller markets and individual retail clients (Hume, 2019). There are over 2000 businesses moving around 30,000 t of food a day (Hume, 2019). Table 8.2 shows data from a first attempt to know direct food losses during the transportation of fruit and vegetables to the CACM gathered during September 2018, as well as other information related to these losses.

TABLE 8.2 Summary of Losses Recorded and Declared by the Administrators and Transporters Surveyed at the Central de Abastos of Mexico City (CACM).

Product	Origin	Transportation time (h)	Care-during transport	Resupply time (days)	Loss	
					Food loss in weight (t)	Money (MX$)
Oranges	Oaxaca	8	NC[1]	7	35	5300
Papaya	México	2	Use of wooden boxes	15	25	7800
Onions	México	3	NC	7	24	10,000
Potatoes	Hidalgo	5	NC	7	35	8000
Potatoes	Hidalgo	3	NC	4	25	7000

TABLE 8.2   *(Continued)*

Product	Origin	Transportation time (h)	Care-during transport	Resupply time (days)	Loss	
					Food loss in weight (t)	Money (MX$)
Potatoes	México	3	NC	7	25	7000
Onions	México	7	NC	7	15	9000
Herbs[2]	Hidalgo, México	6	Avoiding stacking product[3]	3	50	5000
Strawberries/ Blueberries	México	4	Use of plastic boxes	5	40	6000
Tangerine	Oaxaca	8	NC	15	40	4000
Grapefruit	Oaxaca	9	NC	7	25	8000
Oranges	Veracruz	11	NC	10	30	9000
Lemons	México	4	NC	7	25	3000
Mangoes	Oaxaca	5	Use of wooden boxes	7	30	8000
Herbs	Oaxaca, México	11	Use of wooden boxes	3	25	9000
Herbs	México	4	Avoiding stacking product	3	25.	8000
Mangoes	Oaxaca	7	Avoiding stacking product	8	25	6000
Onions	México	4	Use of wooden boxes	7	25	7000
Herbs	México	3	Avoiding stacking product	3	30	10,000
Jicama	Hidalgo	4	NC	30	15	3000

[1]They do not have any care with the food that is transported by truck.
[2]Herbs comprise epazote, parsley, coriander, peppermint, and chamomile.
[3]Avoiding to place a large amount of weight on the fruit or vegetable.

Data in Table 8.2 reveal the large amount of fruit and vegetable losses that we registered at the arrival to the CACM during the time the survey was carried out (569–81 t/day). There is scarce information of the food losses in the CACM, particularly, about the food losses during its transportation from the farm to the wholesale market. However, there are some studies of the total solid waste generated in the wholesale market of Mexico City (Robles-Martínez et al., 2010; Morales-Pérez, 2011) and a recent journalistic chronicle that verifies the accumulation of trash observed in this wholesale market (Hume, 2019). In this regard, Morales-Pérez (2011) estimated that solid waste produced in the CACM corresponded to 503.6 t/day (50% came from the fruit and vegetable zone). This value is similar to the estimated values reported by Fehr et al. (2002) in a Brazilian wholesale market and demonstrates the huge loss of fruits and vegetables in the CACM. According to the National Institute for Federalism and Municipal Development of Mexico (INAFED), a wholesale market is "a commercial unit for the distribution of food products that provides the population with services for the supply of basic products to the wholesale, through facilities that allow the concentration of products from different production centers, and then supply them to the retailers" (INAFED, 2010). The INAFED also defines that the main activities of a central supply are the reception, exhibition, and specialized storage, as well as the sale of food products. In this context, a wholesale market should reduce the inappropriate handling of products and therefore reduce food losses. However, the amount of solid waste (mostly made up of food losses) that wholesale markets generate is large (Riancho et al., 2002; Robles-Martínez et al., 2010; Morales-Pérez, 2011).

As stated in Section 8.2, it is interesting to note that in general, the studies assessing the losses of fruit and vegetables (or other types of food) at this point of the food chain (see Table 8.1) consider food losses at equal to zero (or almost negligible). This is because on one side surveyed farmers transport food to the nearest wholesale market implying that the distance food travels is small. On the other hand, transportation loss values are not disaggregated so that it is not possible to know the specific loss due to the transportation from the farm to the wholesale markets. In the first case, Table 8.2 shows that some fruits and vegetables are not only transported from the cities in which they are produced (55%) but also 45% come from other markets in Mexico City (see e.g., papaya and onions from Mexico). In two cases, administrators reported that herbs were acquired from two different origins. The factors that may influence the food loss of the studied vegetables and fruits will be discussed subsequently (Section 8.4.1).

8.4.1 FACTORS THAT MAY INFLUENCE FOOD LOSSES REGISTERED IN THE CACM DURING ITS TRANSPORTATION FROM THE FARM AND OTHER MARKETS

The surveys applied to the transporters of food and administrators of the stores visited in the CACM allowed us to obtain information about time of transportation of the fruit and vegetables studied; the use of refrigeration during its transportation; and the care given to food by the transporters during the transportation. These surveys also gave us additional information about other factors that may affect food losses such as the use of food considered as loss by the administrators, the reasons given by administrators for food loss, the ownership of trucks facilitated in the transportation of this food, the time in which administrators buy fruit and vegetables again (resupply time), and the amount of money lost due during transportation.

Statistical analysis revealed that there were no differences in the loss of fruit and vegetables ($p > 0.05$). ANOVA test showed that the transportation time between Oaxaca and Mexico City as well as between Mexico City and the cases where food was purchased in two different states were significantly different ($p < 0.05$). In general, despite the difference in these distances, there were no differences among food loss of the different states registered, even though, the loss assessed from trucks coming from Oaxaca and two different states were slightly higher.

The type of care declared by transporters did not affect the amount of food loss ($p > 0.05$) suggesting that it was not sufficient even though they considered it. This poor care was confirmed by the lack of use of refrigeration during the transport of the food analyzed (see Table 8.2). As discussed in Section 8.2, enhancement of food loss due to the lack of a cold chain has been already highlighted by Parfitt et al. (2010) and Kitinoja (2013). Thus, these findings suggest that it is the lack of refrigeration in the trucks transporting the food, as well as the type of care given during the transportation, which are the key factors influencing food loss during the transportation of the sample of fruit and vegetables studied from farms and other markets to the CACM.

8.4.2 GENERAL PERCEPTION OF THE ADMINISTRATORS OF THE CACM OF THE FRUIT AND VEGETABLE LOSS DURING COMMERCIALIZATION

Some of the questions made to the administrators of the CACM show their general perception about the loss of the fruit and vegetables during its commercialization. Surveys revealed that none of the managers interviewed revealed knowing why they lost food during commercialization. Only four mentioned being careful with the food at the wholesale market and only one declared taking advantage of the losses of jicama, a product that registered lower losses when it arrived at the CACM when compared to the other products analyzed (Table 8.2). We assessed the estimated economic losses from three different sides:

- The origin of transported food
- The amount of registered loss depending on each origin
- The resupply time

Interestingly, different tendencies between the estimated economic loss of the administrators and the amount of food loss depending on the origin of the food were observed. For example, when food came from Oaxaca and Hidalgo, high and significant relationships were observed, even though the first case was negative and the second case was positive. In the case of Mexico City, a low relationship was assessed. These tendencies are in accordance with the resupply time: highly related in the cases of Oaxaca and Hidalgo, and poorly related in the case of Mexico City. Moreover, there were no differences among the estimated economic losses and the different origins of transported food ($p > 0.05$). These results may suggest that the economic estimation of fruit and vegetables coming from Mexico City is more complicated or influenced by other factors by administrators. We should mention that we did not investigate how administrators calculated these losses. These results show the complexity of the perception of food loss by the administrators of the CACM. An analysis of the economic impact of food loss in consumers will be made in Section 8.2.

According to Morales-Pérez (2011), in the past decade, there have been mechanisms to take advantage of the losses in the CACM through food banks and other social institutions. Food banks are private assistance institutions that are dedicated to the collection of food loss to bring them in good condition to marginalized families in different states of the country (Banco de Alimentos Caritas del estado de México, 2017). Donating the food loss does not only generate a cost for the tenant or administrator, but it also permits them to deduct an amount equivalent to 5% of the cost of the sale that would have corresponded to the goods that were donated (SAT, 2012). Therefore, the fact that administrators in CACM do not know what to do with fruit and vegetable losses suggests that neither everyone knows they can donate lost food, nor are they interested in donating them. Thus, authorities may promote the donation of food as a strategy to reduce food loss in the CACM.

8.5 THE ECONOMIC IMPACT OF FOOD WASTE AND FOOD LOSS

International studies conclude that food waste and food loss at the world level are equivalent to one-third of the total food output, which amount to 1.3 billion t/year (Gustavsson et al., 2013); this waste has an estimated value of one billion U.S. dollars (Gustavsson et al., 2013). In Mexico, wasted food amounts to 37.4% of total output, which equals 20.4 million tons, with a value of 24.5 million U.S. dollars at market price, slightly more than 2% of the Mexican GDP (Aguilar Gutierrez, 2019; World Bank, 2017).[1]

An important variable in any modern economy is the inflation rate. The existence of food loss and food waste (FLW) is one of the explaining variables of that inflation rate, due to the monetary cost of this loss being transferred to the next step of the chain toward the end consumers' families. Food waste influences on the ability of different social strata to access food. Food pricing is influenced not only by economic factors, but also by noneconomic factors, which result in families, particularly of low-income deciles, not having access to adequate nutrition.[2]

Enhancement of human nutrition and food security depends not only on production but also on distribution, manufacture, and changes in consumption habits. Demand for foodstuff, given that consumer preferences change at a slow rate, is mainly determined by population growth and by changes in income levels and its distribution. In low-income countries, almost any improvement in income levels is followed by fluctuations in pricing of foodstuffs, which have a negative impact on the consumer's real income, lowering aggregate demand, and preventing its modification toward more technologically advanced goods. Food waste increases, mainly, distribution costs, which if reduced could reduce the previously commented effect.

---

[1]Aguilar's initial study (written in 2016 but published in 2019) developed a methodology for calculating the magnitude of food losses and food waste in 2014: that study considered a set of 34 foods from the basic diet of Mexicans, and quantified an approximate loss of 35% (about 11 million tons of waste per year). However, this study was followed and expanded by the World Bank in Mexico in 2017, with data from 2016: it found that for a set of 79 products, the waste is approximately 34% (20.4 million tons of food thrown away annually, in Mexico. Both studies approximated an economic calculation that is around 2% of the gross domestic product, in both years.

[2]An exhaustive analysis of the nature, origin, and causes of waste would lead to a strict formulation of political economy of food waste; situation that was developed by Aguilar (2020).

Estimates of FLW include the quantification of production, transport,[3] and commercialization costs. On one hand, production costs are considered with transportation costs to wholesale centers and then to retail, and finally to the end consumer. These estimates landed on the conclusion that food waste and food loss in Mexico stands at about 25,000 million U.S. dollars per year.[4]

Food consumption problems have incidence on development opportunities of families and nations, signaling the need of government intervention to promote a sustained development. The deployment of innovations that reduce food waste can enhance consumer welfare and social development. This research attempts to contribute in said search.

Within a neoclassical point of view, FLW modifies the market equilibrium through changes in both consumer and supplier surplus.

Consumer surplus is the difference between the price the consumer is willing to pay for a product as a limit and the market price is what he actually pays. If market price is below his limit the consumer can be persuaded to buy more, increasing the quantity demand or the consumer surplus. Thus, any reduction in FLW will increase consumer surplus and demand for other goods.

Supplier surplus, on the other hand, is the difference between the lowest price the manufacturer is willing to accept before not selling a product and the price he actually receives in the market. This surplus is part of his economic gain of conducting business. FLW, by diminishing available supply, can increase supplier surplus in basic foods, which tend to have a very low price elasticity, enhancing supplier surplus, and monetary gains.

Consumer and supplier surplus analysis constitute a way of analyzing the effect of FLW upon market equilibrium. Though not a part of this study, we will always consider that FLW has a negative impact upon consumer surplus and a positive influence upon supplier and trader surplus.

---

[3]Food transportation costs can severely impact the price of food and transportation efficiency affects the quality and value of fresh food. A study for Slovenia showed that the loss of quality in the transport of vegetables represents a significant cost for companies (Osvald and Sadnik, 2008). In the case of Mexico, the costs associated with food transportations are very high, as fresh food must travel hundreds of kilometers: on average, tomato cargo travels 484 km to the nearest wholesale centers, but this distance can be 2838 km in worse instances (COFECE, 2013).

[4]Strictly speaking, the total quantification of the economic impact of food losses and food waste is broader: it must consider not only the costs related to the disbursement made to produce, transport, and distribute food that, for logistical or market reasons, ends up in the garbage. In addition to this quantification, the economic cost associated with carbon dioxide emissions into the atmosphere and the water used to grow food that ends up being lost or wasted should be added (i.e., the environmental cost also can be quantified economically). Therefore, the calculation referred to her, in any circumstance is conservative, in relation to the true economic value of food losses and food waste in Mexico.

8.6 CONCLUSION

A sample of fruits and vegetables marketed at the Mexico City's wholesale market revealed that a significant amount of food products are lost during transport. In general, administrators of local stores at CACM reported having little care in the transportation and during the time they stay in the CACM before its selling to other retailers. Although there are no direct data on the quality of fruits purchased at the sites of origin, the fact that the variable related to the type of food (fruit or vegetable) had not been significant, as well as the fact that the type of care declared by transporters did not affect the amount of food loss demonstrate the importance of proper transport and handling to reduce food losses. At an economic level, food loss increase affects food cost thus affecting social development, thus, reduction of food loss should be promoted institutionally. There is no simple strategy to reduce food loss. However, the findings of the present study suggest that at least at the wholesale level, campaigns to identify what food can be donated, as well as the donation itself, can be given to CACM administrators.

At the end, the reduction of food loss is the task of all of us for our own sake and future generations.

ACKNOWLEDGMENTS

The authors thank Mr. William Bevalet for his support with the revision of the grammar of the present chapter. Also, the authors thank Prof. Juan Antonio Villanueva Jiménez and Dr. Ezequiel Arvizu Barrón for their valuable comments.

KEYWORDS

- **central market**
- **food loss**
- **transportation**
- **economic impact**
- **scientific community**

REFERENCES

Aguilar Gutierrez, G. Responsabilidad Social Corporativa en las Pérdidas y Desperdicios de Alimentos en México. *Brazil. J. Latin Am. Stud.* **2019,** *17*, 168–197.

Aguilar, G. G. *Hambre y Desperdicio en México, 1era. ed.*; Miguel Angel Porrúa, CONACYT and IPN: Ciudad de México, 2020.

Bagherzadeh, M.; Inamura, M.; Jeong, H. *Food Waste along the Food Chain*; OECD Food, Agriculture and Fisheries Papers, No. 71; OECD: Paris, 2014.

Banco de Alimentos Caritas del estado de México. http://bamex.org/ (July 25, 2017).

Beretta, C.; Stoessel, F.; Baier, U.; Hellweg, S. Quantifying Food Losses and the Potential for Reduction in Switzerland. *J. Waste Manage.* **2013,** *33*, 764–773.

Blond, R. D. *The Agricultural Development Systems Project in Egypt*; University of California: Davis, CA, 1984.

Braudel, F. *Civilisation matérielle, économie et capitalisme: XVe-XVIIIe siècle*; Colin: Paris, 1979.

Cadilhon, J. J.; Fearne, A. P.; Hughes, D. R.; Moustier, P. *Wholesale Markets and Food Distribution in Europe: New Strategies for Old Functions*; Imperial College: Wye, London, 2003.

Chaboud, G. Assessing Food Losses and Waste with a Methodological Framework: Insights from a Case Study. *Resour. Conserv. Recycl.* **2017,** *125*, 188–197.

COFECE. Reporte Sobre las Condiciones de Competencia en el Sector Agroalimentario; COFEE: Ciudad de México, Mexico, 2013.

FAO (Organización de las Naciones Unidas para la Agricultura y la Alimentación). Losses and Food Waste in Latin America and the Caribbean. http://www.fao.org/americas/noticias/ver/en/c/239392/ (September 10, 2018).

Fehr, M.; Calcado, M. D. R.; Romao, D. C. The Basis of a Policy for Minimizing and Recycling Food Waste. *Environ. Sci. Policy.* **2002,** *5* (3), 247–253.

Gustavsson, J.; Cederberg, C.; Sonesson, U.; Emanuelsson, A. The Methodology of the FAO Study: Global Food Losses and Food Waste-Extent, Causes and Prevention—FAO, 2011; SIK report No. 857; SIK-The Swedish Institute for Food and Biotechnology: Sweden, 2013.

Hume, A. Central de Abasto, Mexico City's Wholesale Market, Is a City unto Itself. *Mexico News Daily* [Online] July 17, 2019. https://mexiconewsdaily.com/mexicolife/mexico-citys-wholesale-market/ (accessed December 02, 2019).

INAFED (Instituto Nacional Para el Federalismo y el Desarrollo Municipal). *Guía técnica 14. La Administración de Mercados y Centrales de Abasto.* http://www.inafed.gob.mx/work/models/inafed/Resource/335/1/images/guia14_a_administracion_de_mercados_y_centrales_de_abastos.pdf (October 21, 2010).

Kitinoja, L.; Al-Hassan, H. Y. In Postharvest Technology in the Global Market, Proceedings of the *XXVIII International Horticultural Congress on Science and Horticulture for People*, Lisbon, Portugal, August 31-40, 2010; Cantwell, M. I.; Almeyda D. P. F., Eds., Lisbon, 2012.

Kitinoja, L. Use of Cold Chains for Reducing Food Losses in Developing Countries. *Population* **2013,** *6*, 5–60.

Osvald, A.; Zadnik Stirn, L. A Vehicle Routing Algorithm for the Distribution of Fresh Vegetables and Similar Perishable Food. *J. Food Eng.* **2008,** *85*, 285–295.

Saran, S.; Roy, S. K.; Kitinoja, L. Appropriate Postharvest Technologies for Improving Market Access and Incomes for Small Horticultural Farmers in Sub-Saharan Africa and

South Asia. Part 2: Field Trial Results and Identification of Research Needs for Selected Crops. *Acta Hortic.* **2012**, *934*, 41–52.

Kling, W. Food Waste in Distribution and Use. *J. Farm Econ.* **1943**, *25* (4), 848–859.

Koester, U.; Empen, J.; Holm, T. Food Losses and Waste in Europe and Central Asia. *Draft Synthesis Report*; FAO: Rome, Italy, 2013.

Koester, U. Food Loss and Waste as an Economic and Policy Problem. *Intereconomics* **2014**, *49* (6), 348–354.

Jedermann, R.; Nicometo, M.; Uysal, I.; Lang, W. Reducing Food Losses by Intelligent Food Logistics. *Phil. Trans. R. Soc. A* **2014**, *372*, 2013030.

Linares, E.; Bye, R. Traditional Markets in Mesoamerica: A Mosaic of History and Traditions. In *Ethnobotany of Mexico;* Lira, R., Casas, A., Blancas, J., Ed.; Springer: New York, 2016; pp. 151–177.

Morales-Pérez, R. E. Planes de manejo de residuos de generadores de alto volumen: El caso de la Central de Abasto del Distrito Federal, México. PhD Dissertation. Instituto Politécnico Nacional. CDMX, México, 2011.

Parfitt, J.; Barthel, M.; Macnaughton, S. Food Waste within Food Supply Chains: Quantification and Potential for Change to 2050. *Philos. Trans. R. Soc. Lond. B: Biol. Sci.* **2010**, *365* (1554), 3065–3081.

Prusky, D. Reduction of the Incidence of Postharvest Quality Losses, and Future Prospects. *Food Secur.* **2011**, *3*, 463–474.

Robles-Martínez, F.; Ramírez-Sánchez, I. M.; Piña-Guzmán, A. B.; Colomer-Mendoza, F. J. Efecto De La Adición De Agentes Estructurantes a Residuos Hortícolas en Tratamientos Aerobios. *INAGBI* **2010**, *2* (1), 45–51.

SAT (Servicio de Administración Tributaria). http://m.sat.gob.mx/fichas_tematicas/donacion_mercancias_banco_alimentos/Paginas/decreto_beneficios_fiscales.aspx (accessed mes (ingles) día (número), 2012).

Riancho, M. R. S.; Nájera Aguilar, H. A.; Ramírez Herrera, J. G.; Mejía Sánchez, G. M. Aplicación Del Composteo Como Método De Tratamiento De Los Residuos De Frutas Producidos En Zonas De Alta Generación. *Ingeniería* **2002**, *6* (1), 13–20.

Taboada-González, P.; Aguilar-Virgen, Q.; Cruz-Sotelo, S. E.; Ramírez-Barreto, M. E. Manejo y Potencial de Recuperación de Residuos Sólidos en Una Comunidad Rural de México. *Rev. Int. Contam. Ambie* **2013**, *29*, 43–48.

Tracey-White, J. D. *Wholesale Markets: Planning and Design Manual*; Tracey-White, J. D., Ed.; FAO Agricultural Service Bulletin, Series 90; FAO: Rome, 1991.

Villamar, A. A. El estudio etnobioecológico de los tianguis y mercados en México. *Etnobiología* **2016**, *14* (2), 38–46.

World Bank. Pérdidas y Desperdicios de Alimentos en México. Una Perspectiva Económica, Ambiental y Social; Unidad de Desarrollo Sostenible, Región de América Latina y el Caribe, Grupo Banco Mundial: 2017.

CHAPTER 9

Cereal By-products as Source of Dietary Fiber

ERICK P. GUTIÉRREZ-GRIJALVA[1], LUIS A. CABANILLAS-BOJÓRQUEZ[2], GABRIELA VAZQUEZ-OLIVO[2], and J. BASILIO HEREDIA[2*]

[1]*Cátedras CONACYT-Centro de Investigación en Alimentación y Desarrollo A.C., Carretera a Eldorado Km. 5.5, Col Campo El Diez, Culiacán, Sinaloa 80110, México*

[2]*Centro de Investigación en Alimentación y Desarrollo A.C., Carretera a Eldorado Km. 5.5, Col Campo El Diez, Culiacán, Sinaloa 80110, México*

[*]*Corresponding author. E-mail: jbheredia@ciad.mx*

ABSTRACT

Cereals are the most cultivated crops around the world with a total production of around 2,723,878,753 t by 2017. During the processing of cereals, most industries only use grains for human and animal feeding purposes, and the rest of the plant, namely stalks, husk, and leaves, are usually considered as waste. Agro-industrial waste is not usually adequately treated and discarded. Thus, it is an important source of contamination. As an attempt to alleviate food waste contamination, several studies have been conducted to find an economically profitable use of this waste. Most studies have pointed out that cereal by-products are the main source of nutraceutical ingredients such as dietary fiber constituents like lignin, hemicellulose, and cellulose, among others. The components of dietary fiber have attracted much attention because they are the majoritarian compounds found in food waste and due to their reported prebiotic properties. This chapter aims to comprehensively review the latest reports on the bioactive potential of dietary fiber components obtained from cereal by-products.

9.1 INTRODUCTION

Cereals have been an important food in the human diet for a long time. The term "cereals" is referred to the plants from the *Poaceae* family (also known as *Gramineae*), grown for its edible seeds (grains). The most produced grains are wheat, rice, and maize.[1] As reported by the Food and Agriculture Organization, the world production of cereals in 2017 was 2980.174 million tons. The major cereals produced in the world in 2017 were corn (maize) (1134.74 MT), wheat (771.71 MT), rice (769.65 MT), barley (147.40 MT), sorghum (57.60 MT), millet (28.45 MT), oat (24.94 MT), and rye (13.73 MT) (Fig. 9.1).[2]

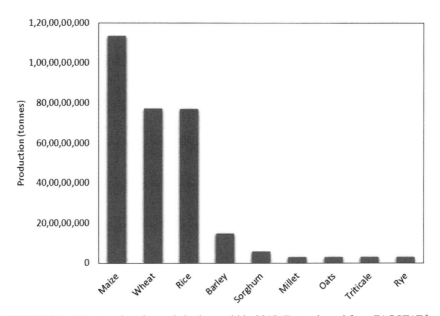

FIGURE 9.1   Most produced cereals in the world in 2017. Data adapted from FAOSTAT.[2]

The main by-products from cereal milling are bran and germ. These are discarded because they affect the properties of flour.[3] The by-products obtained depend on the type of cereal and the type of processing. Overall, cereal by-products are derived from the dry and wet milling and brewing. Dry milling separates the starchy endosperm from the outer layers of the grain, whereas wet milling separates the germ and bran to produce gluten and starch.

- *Maize*: Maize can be processed by dry and wet milling. Maize bran is generated from dry milling and wet milling. The by-products are maize germ oil meal, maize gluten feed, maize gluten meal, and maize steep liquor.[4]
- *Rice*: The milling process for rice is mainly used to produce white rice, where the grain is dehusked and polished. Therefore, the main by-products from rice processing are rice bran and also broken rice.[5]
- *Wheat*: Wheat grain can be used for the distillery industry, and the by-products from this process are distiller's grains, also, from the brewery industry, the brewer's grain from wheat is generated. Another method for wheat grain is the milling to produce flour, this is obtained by roller milling, and from this process, the germ and external layers of wheat grains are the main by-products.[6]
- *Barley*: Barley is also processed by rolling milling and by pearling. The former generates the seedcoat, while the second removes the pericarp, aleurone, subaleurone, and germ. Barley is also used for malting and brewing, producing bagasse malt, malt sprout, or spent barley grains.[7]
- *Sorghum*: From sorghum, the seed husk is the main by-product. Other by-products include gluten feed from wet milling, and from dry milling, germ, and gluten meal are generated. Also, in some countries, sorghum is used in malting, and the by-product of this process is called sprout. On the other hand, from brewing, the sorghum brewer's spent grain is generated.[8] Furthermore, distillery residues from the production of Kaoliang liquor, a fermented beverage from sorghum, are generated.[9]
- *Oats*: Oats are thermally treated before milling into flour. Oat bran is the main by-product.[10]
- *Rye*: Rye grains are processed by roller milling to obtain the flour. The bran is the main by-product.[10]
- *Millet*: Millets are processed by dry milling to produce flour; therefore, the millet by-products include the endosperm, germ, and bran.[11]

9.2 CEREAL BY-PRODUCTS AS SOURCE OF DIETARY FIBER

Dietary fiber is one of the major components of cereal by-products; it is mainly present in cereal bran. Dietary fiber can be defined as nondigestive dietary constituents that cannot be hydrolyzed by human digestive enzymes.[12] Dietary fiber can be classified based on its water solubility in soluble and insoluble dietary fiber. All cereals have different proportions of soluble–insoluble dietary fiber and are summarized in Figure 9.2.

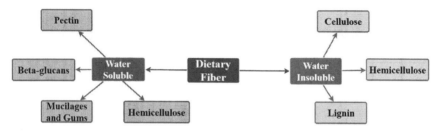

FIGURE 9.2 Types of dietary fiber depending on its water-solubility properties.

- Soluble dietary fiber has a high water-holding capacity and is composed of nonstarch polysaccharides such as β-glucans and soluble hemicelluloses such as arabinoxylans. β-Glucans from cereal brans are polysaccharides of glucose residues containing the β-(1–4) and (1–3) linkages.[13,14] Arabinoxylans are nonstarch structural polysaccharides from the cell wall of cereal grains.[15]
- Insoluble fiber does not have water-holding capacity and is made up of cellulose, hemicellulose, and lignin.[16] Cellulose is the most abundant polysaccharide on earth; it is made of several glucose units with β-(1–4)-glucosidic linkages. Hemicelluloses are heteropolysaccharides made up mainly by xyloses, and in less abundance by other sugars such as arabinose, mannose, and galactose. Pectin is another component of the plant cell wall and is characterized for being rich in galacturonic acid.[17]

TABLE 9.1 Dietary Fiber Distribution in the Main Cereal By-products.

Source of dietary fiber	Total dietary fiber (% dry matter)	Insoluble dietary fiber	Soluble dietary fiber	Reference
Corn bran	54.1	48.1	6.0	Zhao[18]
Wheat bran	47.1	44.8	2.4	Talukder[19]
Rice bran	32.9	30.2	2.7	Daou[20]
Barley bran	17.1		9.2	Karimi[21]
Sorghum bran	66.43	58.17	8.26	Buitimea-Cantúa[22]
Millet bran	79.22	77.57	1.75	Zhu[23]
Oat bran	16.5	8.5	7.9	Talukder[19]
Rye bran	56.10	43.20	5.01	Kajala[24]

Lignin is not a carbohydrate but is a complex heteropolymer made up of phenolic monomers. Lignin is the second most abundant biomolecule in nature, only after cellulose. Physiologically, lignin forms a complex linkage network with lignocellulosic fibers providing rigidity to plant cell walls.[25] Lignin structure is complex, and it contains hydroxy, methoxy, carbonyl, and carboxyl radicals distributed in the molecule.[25] The conformational units of lignin are called monolignols, and the main constituents are p-coumaryl alcohol, coniferyl alcohol, and sinapyl alcohol (Fig. 9.3).[25,26] Gums and mucilage are not cell wall components; however, some are considered as dietary fiber; the former is secreted by the plant in response to plant injury, whereas mucilage is produced to prevent desiccation of seed endosperm.[27]

a)   b)   c)

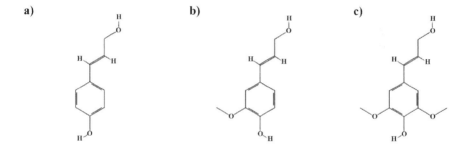

FIGURE 9.3   Lignin composition is made of the monolignols (a) p-coumaryl alcohol, (b) coniferyl alcohol, and (c) sinapyl alcohol.

As aforementioned, dietary fiber is the main constituent of the most common cereal by-product, the bran. Dietary fiber has been related to the prevention of noncommunicable diseases such as cardiovascular diseases, diabetes, obesity, and colorectal cancer.[28] These represent the main cause of death worldwide and are responsible for nearly 70% of all deaths in the world.[29] Moreover, due to the increased interest in healthier foods, market trends for dietary fiber-rich products is on the rise, most consumers are aware of the importance of fiber intake, and new products are there is a 15% annual growth of new products rich in dietary fiber.[30]

One of the most popular products that have been subjected to study in the valorization of cereal by-products, like dietary fiber from cereal bran, are bakery products.[31] Nonetheless, cereal bran does not only contain soluble and insoluble dietary fiber, but also proteins, minerals, and in some cases, some antinutrient components like phytic acid, which may be degraded after some technological approaches.[31–33]

As the interest in dietary fiber from cereal bran grows, new processing techniques have been studied. Some can be found summarized in Luithui[14] and Saleh.[34]

9.2.1 DIETARY FIBER FROM CEREAL BY-PRODUCTS

As evidenced in Table 9.1, the main by-product of cereals is the bran, and the main constituent of bran is the soluble and insoluble dietary fiber. Thus, most reports are characterization and valorization studies of the types of fibers in each cereal by-product, their potential bioactive properties, and potential as a functional ingredient. Dietary fiber from cereal by-products is of interest in the dietary supplement and food industries. Dietary supplements from plant-based ingredients have been an ongoing trend in the food market due to the rising interest of consumers for healthier food.[30]

Bran composition from the different cereals may be different depending on plant species and variety. Thus, cereals will have a different ratio of soluble–insoluble dietary.[28] Dietary fiber consumption is related to many health benefits in humans, and these benefits are dependent on the type of fiber. For instance, soluble fiber is related to reduced cardiovascular diseases, lowered plasma cholesterol, and blood sugar levels.[14]

Arabinoxylans are the most abundant group of constituents that have been subject to study in xylans from the bran of different cereals.[35,36] The most common type of arabinoxylans in cereals are β-ᴅ-xylopyranoside units and sometimes linked with α-ʟ-arabinofuranoside, galacturonic acid, 4-O-methylglucoronic acid, ʟ-arabinose, ᴅ-xylose, ᴅ-galactose, ᴅ-glucose, among others.[17,36]

Arabinoxylans have been linked to antioxidant and prebiotic properties by in-vitro and in-vivo studies.[35–37] Herrera-Balandrano et al.[38] evaluated the suitability of inclusion of feruloylated arabinoxylans from nixtamalized maize bran to frankfurter sausages. The authors reported that addition of feruloylated arabinoxylans from nixtamalized maize bran from 0.15% to 0.30% increased the phenolic content and antioxidant capacity of the sausages, which may be attributed to the presence of ferulic acid between arabinose and xylose linages.[38] Feruloylated arabinoxylans in corn and wheat bran have been identified by HPLC and LC–MS/MS as 5-O-feruloyl-α-ʟ-arabinofuronase, and 5-O-feruloyl-α-ʟ-arabinofuranosyl-(1-3)-O-β-ᴅ-xylopyranose (tentative identification).[35]

Corn bran has also been studied by Malunga et al.,[39] who found that feruloylated arabinoxylan mono and oligosaccharides inhibit α-glucosidase and
glucose transporters. It is important to mention that the authors highlighted
that the potential hypoglycemic effect of feruloylated arabinoxylans is
ferulic acid-dependent and managed to inhibit glucose transporter 2.[39] Corn
bran has been studied as a food supplement in snack bars by de Sousa et
al.,[40] and the authors report that the inclusion of corn bran to bars increased
the dietary fiber content and had good acceptability. Similarly, Pontonio[33]
concluded that the fortification of wheat bread with maize by-products
increases dietary fiber and proteins; however, the authors took a different
approach, they used lactic acid fermentation to increase protein digestibility
up to 60%, decreased nearly 13% of starch hydrolysis, and degradation of
phytic acid. Furthermore, the sensory acceptability of the corn-bran-enriched
bread showed a more balanced profile.[33]

Moreover, some studies indicate that arabinoxylans from corn bran may
modulate the immune system.[36] Zhang et al.[41] obtained arabinoxylan oligosaccharides using corn bran treated with endoxylanases. The authors reported
an increased nitric oxide synthesis in monocytes from 53.7 to 62.9 μM per
million viable cells at a concentration of arabinoxylan of 5–1000 μg/mL.
The authors also mention the importance of future research to understand
the effect of arabinoxylan's molecular weight on their immune-modulating
potential.[41]

Rice is the third most cultivated cereal in the world and provides around
two-thirds of the caloric ingestion for about 4.5 billion people. Rice is
commonly consumed dehulled, and rice bran is considered as a by-product[42].
Rice bran is a source of proteins, soluble and insoluble dietary fiber, minerals,
fatty acids, and some phytochemicals. γ-Oryzanol is an ester of the hydroxyl
group of sterols or triterpene alcohols with the carboxylic group of ferulic
acid.[42,43] Recently, de Souza[42] elaborated cookies enriched with rice bran, the
authors reported increased content of bioactive compounds, which increased
the antioxidant capacity of cookies as well as their dietary fiber content. The
highest acceptance and preference were for cookies formulated with 75%
of rice bran. Among the bioactive compounds found in cookies with this
formulation were dietary fiber, lutein, and γ-oryzanol.

Also, Jia[44] evaluated dietary fiber from rice bran, where the fermentation
by *Trichoderma viride* was optimized with surface-methodology response
methodology. Their work aimed to optimize an extraction method of
functional dietary fiber from rice bran to be used as an ingredient in the
development of functional foods. Soluble fiber showed increased yield

during fermentation from 10.5% to 33.4%; moreover, fermented dietary fiber showed increased molecular weight, more complex monosaccharide composition, and better functional properties like cholesterol absorption capacity than nonfermented rice bran fiber.[44]

Dietary fiber from rice bran can be used as an ingredient for producing symbiotic formulations as a study by Saman[5] shows. Their work used different rice bran fractions to support the growth of *Lactobacillus plantarum* NCIMB 8826 and *Lactobacillus reuteri* NCIMB 8821, two known probiotic strains. *L. plantarum* showed better results during fermentation, yielding higher cell biomass; thus, the authors state that "fractionation using controlled debranning might be an efficient and inexpensive way to obtained functional ingredients for the preparation of symbiotic food products."[5]

Furthermore, research suggests that the microbiota metabolism of prebiotics has a relationship in the chemoprevention of colorectal cancer. Thus, the ingestion of foods with high dietary fiber content is encouraged. A randomized-controlled pilot clinical trial developed by Sheflin,[45] administered heat-stabilized rice bran at a concentration of 30 g/day and evaluated its effect on gut microbiota and metabolites, on 29 overweight/obese patients with a clinical history of colorectal cancer. The administration of rice bran increased bacterial diversity after 14 days. However, the authors state that more "research is needed to determine if rice bran changes on microbiota are lasting".[45]

The research by Qi[46] showed that rice bran insoluble dietary fiber has glucose absorption capacity and α-amylase inhibitory capacity. This effect was associated with the formation of a complex between rice bran insoluble dietary fiber and amylase. The authors conclude that the extraction of insoluble dietary fiber using a combination of acid (sulfuric acid) and alkaline (potassium hydroxide) treatments significantly improves the ability of rice bran dietary fiber on its potential hypoglycemic properties.[46]

As mentioned in Table 9.1, wheat bran is a rich source of insoluble dietary fiber. The main constituents of insoluble dietary fiber are cellulose and hemicellulose, which have been reported with a prebiotic activity enhancing the growth of specific gut microbiota populations with health benefits.[27,28] Chen[47] evaluated the effect of fermented dietary fiber from wheat bran on the expression of genes involved in short-chain fatty acid (SCFA) transport, G-protein signaling, apoptosis, cell proliferation, and oncogenesis in the colon epithelium of healthy rats. Fermentable insoluble dietary fiber from wheat bran modulated the expression of the SCFA transporters Mct and Smct2, also modulated the expression of genes involved in G-protein signaling. The authors suggest that this effect might be mediated through the Tp53-dependent SCFA transport mechanism. The study concludes that a systematic approach is needed to elucidate the cellular mechanism behind the antiproliferative potential of wheat bran fermentable dietary fiber.

Resistant starch is a fraction of starch that is resistant to gastrointestinal digestion enzymes; thus, it reaches the large intestine where it can be fermented by gut microbiota.[48,49] Resistant starch from wheat bran, as reported by Kahraman,[49] was incorporated in a cookie formulation at the concentrations of 25%, 50%, and 75% aiming to obtain low glycemic index cookies. The authors report that the addition of resistant starch from wheat bran improved cookie quality, increased dietary fiber content, and lowered in-vitro glycemic index in cookies.

Morevoer, Malunga and Beta[50] evaluated the effect of xylanase from two sources, *T. viride* or *Neocallimastix patriciarum*, and graded ethanol fractionation on the antioxidant capacity of arabinoxylan oligosaccharides from wheat bran. The authors reported that ferulic acid content from arabinoxylans treated with *T. viride* was two-fold higher than the samples treated with *N. patriciarum*. A recent work by Pontonio[32] combined the effect of xylanase treatment and lactic acid bacteria fermentation of wheat by-products from different varieties on their biochemical and nutritional features. Wheat bran total dietary fiber content from the different varieties ranged between 10% and 26.3%, insoluble dietary fiber was the most abundant in all varieties. The authors prepared bread, adding 30% of fermented wheat bran, and the total dietary fiber content of bread ranged between 1.87% and 7.15%. The use of wheat bran as a functional ingredient in breadmaking could be a good strategy to increase the average intake of dietary fiber on the population.

As aforementioned, arabinoxylans from cereal by-products are of interest in the food industry due to their prebiotic potential, thus improving their extraction has been the goal of some studies. Lie Vangsoe,[51] evaluated the enzymatic production of arabinoxylans from wheat bran. The authors studied different cell-wall-degrading enzymes such as xylanase, cellulase, and β-glucanase. The authors report that the arabinose/xylose ratio is a determinant for arabinoxylan yield as they inversely correlate. Furthermore, the use of cell-wall-degrading enzymes yields mainly arabinoxylan oligosaccharides, which have been reported with prebiotic activity.[36,37]

Another study using enzymatic treatments to extract arabinoxylans from wheat bran was reported by Xue,[52] who used xylanase and arabinofuranosidase from *Thermotoga martima*. Enzyme action yields xylooligosaccharides, and inclusion of wheat bran on breadmaking improved the dough rheology and nutritional characteristics of steamed bread. Additionally, the content of phenolic acids and antioxidant capacity of the bread with enzyme-extracted xylooligosaccharides were higher. Xylanase-treated wheat bran yields a total phenolic content of 0.50 mg GAE/g, and arabinofuranosidase-treated had a total phenolic content of 0.45 mg GAE/g.

Yan[53] aimed to use the blasting extrusion process to increase the concentration of soluble dietary fiber from wheat bran, with the purpose to extract purified soluble polysaccharides such as arabinoxylans and feruloyl oligosaccharides. Their study showed that extrusion yield improved the content of soluble dietary fiber from wheat bran without the loss of nutritional compounds. Extruded wheat bran dietary fiber had a high content of arabinose, xylose, and glucose, and low content of galactose, with molar ratios of 0.76:0.99:1:0.12.[53]

9.3   CONCLUSION

The main by-product of cereal processing is cereal bran. This by-product is rich in dietary fiber and can be used as a functional ingredient in the food industry. The awareness of customers for healthier food choices make dietary fiber-rich foods a target product for many people. Recent studies have focused on new strategies to obtain bran dietary fiber constituents such as oligosaccharides, which have prebiotic activity, antidiabetic potential, and might prevent the onset of colorectal cancer. Further studies are needed to elucidate the underlying mechanisms of the health benefits of dietary fiber from cereal by-products like cereal bran. Cereal by-products are a safe choice for functional ingredients after treatment to eliminate antinutrient constituents.

KEYWORDS

- **cereal by-products**
- **dietary fiber**
- **agro-industrial waste**
- **source of contamination**
- **bioactive**

REFERENCES

1. Tull, A. Food Commodities. In *Home Economics OCR Food and Nutrition for GCSE*; Tull, A., Ed.; Hodder Education: Italy, 2009; p 115.

2. FAOSTAT. *Food and Agriculture Data*. http://www.fao.org/faostat/es/#data/QC (accessed December 20th, 2019).

3. Verni, M.; Rizzello, C. G.; Coda, R. Fermentation Biotechnology Applied to Cereal Industry By-products: Nutritional and Functional Insights. *Front. Nutr.* **2019,** *6* (42), 1–13.

4. Amado, I. R.; Vázquez, J. A.; Pastrana, L.; Teixeira, J. A. Microbial Production of Hyaluronic Acid from Agro-Industrial By-products: Molasses and Corn Steep Liquor. *Biochem. Eng. J.* **2017,** *117*, 181–187.

5. Saman, P.; Fuciños, P.; Vázquez, J. A.; Pandiella, S. S. By-products of the Rice Processing Obtained by Controlled Debranning as Substrates for the Production of Probiotic Bacteria. *Innovative Food Sci. Emerging Technol.* **2019,** *51*, 167–176.

6. Brandolini, A.; Hidalgo, A., Wheat Germ: Not Only a By-product. *Int. J. Food Sci. Nutr.* **2012,** *63* (suppl 1), 71–74.

7. Nigam, P. S. An Overview: Recycling of Solid Barley Waste Generated as a By-product in Distillery and Brewery. *Waste Manage.* **2017,** *62*, 255–261.

8. Ratnavathi, C. V.; Chavan, U. D. Chapter 2Malting and Brewing of Sorghum. In *Sorghum Biochemistry*; Ratnavathi, C. V., Patil, J. V., Chavan, U. D., Eds.; Academic Press: San Diego, 2016; pp 63–105.

9. Wang, C.-Y.; Ng, C.-C.; Lin, H.-T.; Shyu, Y.-T. Free Radical-Scavenging and Tyrosinase-Inhibiting Activities of Extracts from Sorghum Distillery Residue. *J. Biosci. Bioeng.* **2011,** *111* (5), 554–556.

10. J. Sibakov, P. L. a. K. P., Cereal Brans as Dietary Fiber Ingredients. In *Fibre-Rich and Wholegrain Foods: Improving Quality*; Delcour, J. A., Poutanen, K., Eds.; Elsevier: USA, 2013; pp 176–177.

11. Liu, J.; Tang, X.; Zhang, Y.; Zhao, W. Determination of the Volatile Composition in Brown Millet, Milled Millet and Millet Bran by Gas Chromatography/Mass Spectrometry. *Molecules* **2012,** *17* (3), 2271–2282.

12. Dahl, W. J.; Stewart, M. L. Position of the Academy of Nutrition and Dietetics: Health Implications of Dietary Fiber. *J. Acad. Nutr. Dietetics* **2015,** *115* (11), 1861–1870.

13. Zhu, F.; Du, B.; Xu, B. A Critical Review on Production and Industrial Applications of Beta-Glucans. *Food Hydrocolloids* **2016,** *52*, 275–288.

14. Luithui, Y.; Nisha, R. B.; Meera, M. Cereal By-products as an Important Functional Ingredient: Effect of Processing. *J. Food Sci. Technol.* **2019,** *56* (1), 1–11.

15. Khan, M. A.; Nadeem, M.; Rakha, A.; Shakoor, S.; Shehzad, A.; Khan, M. R. Structural Characterization of Oat Bran $(1{\rightarrow}3)$, $(1{\rightarrow}4)$-β-D-Glucans by Lichenase Hydrolysis Through High-Performance Anion Exchange Chromatography with Pulsed Amperometric Detection. *Int. J. Food Properties* **2016,** *19* (4), 929–935.

16. Sidhu, J. S.; Kabir, Y.; Huffman, F. G. Functional Foods from Cereal Grains. *Int. J. Food Prop.* **2007,** *10* (2), 231–244.

17. Carpita, N. C.; Ralph, J.; McCann, M. C. The Cell Wall. In *Biochemistry and Molecular Biology of Plants*; Buchanan, B. B., Gruissem, W., Jones, R. L., Eds.; John Wiley & Sons, Ltd: New Jersey, USA, 2015.

18. Zhao, J.; Zhang, G.; Dong, W.; Zhang, Y.; Wang, J.; Liu, L.; Zhang, S. Effects of Dietary Particle Size and Fiber Source on Nutrient Digestibility and Short Chain Fatty Acid Production in Cannulated Growing Pigs. *Anim. Feed Sci. Technol.* **2019,** *258,* 114310.

19. Talukder, S.; Sharma, D. P. Development of Dietary Fiber Rich Chicken Meat Patties Using Wheat and Oat Bran. *J. Food Sci. Technol.* **2010,** *47* (2), 224–229.

20. Daou, C.; Zhang, H. Functional and Physiological Properties of Total, Soluble, and Insoluble Dietary Fibres Derived from Defatted Rice Bran. *J. Food Sci. Technol.* **2014,** *51* (12), 3878–3885.

21. Karimi, R.; Azizi, M. H.; Xu, Q.; Sahari, M. A.; Hamidi, Z. Enzymatic Removal of Starch and Protein During the Extraction of Dietary Fiber from Barley Bran. *J. Cereal Sci.* **2018,** *83,* 259–265.

22. Buitimea-Cantúa, N. E.; de la Rosa-Millán, J. Physicochemical, Textural, and In Vitro Starch Digestion Properties of Nixtamalized Maize Flour and Tortillas Enriched with Sorghum (*Sorghum bicolor* (L.) Moench) Bran. *Cereal Chem.* **2018,** *95* (6), 829–837.

23. Zhu, Y.; Chu, J.; Lu, Z.; Lv, F.; Bie, X.; Zhang, C.; Zhao, H. Physicochemical and Functional Properties of Dietary Fiber from Foxtail Millet (*Setaria italic*) Bran. *J. Cereal Sci.* **2018,** *79,* 456–461.

24. Kajala, I.; Mäkelä, J.; Coda, R.; Shukla, S.; Shi, Q.; Maina, N. H.; Juvonen, R.; Ekholm, P.; Goyal, A.; Tenkanen, M. Rye Bran as Fermentation Matrix Boosts in situ Dextran Production by *Weissella confusa* Compared to Wheat Bran. *Appl. Microbiol. Biotechnol.* **2016,** *100* (8), 3499–3510.

25. Vazquez-Olivo, G.; Criollo-Mendoza, M. S.; Gutierrez-Grijalva, E. P.; Picos-Salas, M. A.; Heredia, J. B. Lignin and Its Derivatives as Antioxidant, Antiviral and Antimicrobial Agents: Applicability in Human Health Promotion. In *Lignin: Biosynthesis, Functions and Economic Significance*; Lu, F., Yue, F., Eds.; Nova Science Publishers, Inc.: New York, USA, 2019; pp 261–280.

26. Vanholme, R.; Cesarino, I.; Rataj, K.; Xiao, Y.; Sundin, L.; Goeminne, G.; Kim, H.; Cross, J.; Morreel, K.; Araujo, P.; Welsh, L.; Haustraete, J.; McClellan, C.; Vanholme, B.; Ralph, J.; Simpson, G. G.; Halpin, C.; Boerjan, W. Caffeoyl Shikimate Esterase (CSE) Is an Enzyme in the Lignin Biosynthetic Pathway in Arabidopsis. *Science* **2013,** *341* (6150), 1103–1106.

27. Dhingra, D.; Michael, M.; Rajput, H.; Patil, R. T. Dietary Fibre in Foods: A Review. *J. Food Sci. Technol.* **2012,** *49* (3), 255–266.

28. Alan, P. A.; Ofelia, R. S.; Patricia, T.; Maribel, R. S. R. Cereal Bran and Wholegrain as a Source of Dietary Fibre: Technological and Health Aspects. *Int. J. Food Sci. Nutr.* **2012,** *63* (7), 882–892.

29. World Health Organization. *Global Health Observatory Data.* http://www.who.int/gho/en/ (accessed January 27, 2018).

30. Innova Market Insights Innova Report: Top Ten Trends for 2019. https://www.innovamarketinsights.com/ (accessed December 10).

31. Martins, Z. E.; Pinho, O.; Ferreira, I. Food Industry By-products Used as Functional Ingredients of Bakery Products. *Trends Food Sci. Technol.* **2017,** *67,* 106–128.

32. Pontonio, E.; Dingeo, C.; Di Cagno, R.; Blandino, M.; Gobbetti, M.; Rizzello, C. G. Brans from Hull-Less Barley, Emmer and Pigmented Wheat Varieties: From By-products to Bread Nutritional Improvers using Selected Lactic Acid Bacteria and Xylanase. *Int. J. Food Microbiol.* **2020,** *313,* 12.

33. Pontonio, E.; Dingeo, C.; Gobbetti, M.; Rizzello, C. G. Maize Milling By-products: From Food Wastes to Functional Ingredients Through Lactic Acid Bacteria Fermentation. *Front. Microbiol.* **2019,** *10,* 14.

34. Saleh, A. S. M.; Wang, P.; Wang, N.; Yang, S.; Xiao, Z. G. Technologies for Enhancement of Bioactive Components and Potential Health Benefits of Cereal and Cereal-Based Foods: Research Advances and Application Challenges. *Crit. Rev. Food Sci. Nutr.* **2019,** *59* (2), 207–227.

35. Malunga, L. N.; Beta, T. Isolation and Identification of Feruloylated Arabinoxylan Mono- and Oligosaccharides from Undigested and Digested Maize and Wheat. *Heliyon* **2016,** *2* (5), 18.

36. Broekaert, W. F.; Courtin, C. M.; Verbeke, K.; Van de Wiele, T.; Verstraete, W.; Delcour, J. A. Prebiotic and Other Health-Related Effects of Cereal-Derived Arabinoxylans, Arabinoxylan-Oligosaccharides, and Xylooligosaccharides. *Crit. Rev. Food Sci. Nutr.* **2011,** *51* (2), 178–194.

37. Saeed, F.; Pasha, I.; Anjum, F. M.; Sultan, M. T. Arabinoxylans and Arabinogalactans: A Comprehensive Treatise. *Crit. Rev. Food Sci. Nutr.* **2011,** *51* (5), 467–476.

38. Herrera-Balandrano, D. D.; Baez-Gonzalez, J. G.; Carvajal-Millan, E.; Mendez-Zamora, G.; Urias-Orona, V.; Amaya-Guerra, C. A.; Nino-Medina, G. Feruloylated Arabinoxylans from Nixtamalized Maize Bran Byproduct: A Functional Ingredient in Frankfurter Sausages. *Molecules* **2019,** *24* (11), 11.

39. Malunga, L. N.; Eck, P.; Beta, T. Inhibition of Intestinal alpha-Glucosidase and Glucose Absorption by Feruloylated Arabinoxylan Mono- and Oligosaccharides from Corn Bran and Wheat Aleurone. *J. Nutr. Metabol.* **2016,** *2016,* 9.

40. de Sousa, M. F.; Guimaraes, R. M.; Araujo, M. D.; Barcelos, K. R.; Carneiro, N. S.; Lima, D. S.; Dos Santos, D. C.; Batista, K. D.; Fernandes, K. F.; Lima, M.; Egea, M. B. Characterization of Corn (*Zea mays* L.) Bran as a New Food Ingredient for Snack Bars. *LWT-Food Sci. Technol.* **2019,** *101,* 812–818.

41. Zhang, Z. X.; Smith, C.; Li, W. L.; Ashworth, J. Characterization of Nitric Oxide Modulatory Activities of Alkaline-Extracted and Enzymatic-Modified Arabinoxylans from Corn Bran in Cultured Human Monocytes. *J. Agric. Food Chem.* **2016,** *64* (43), 8128–8137.

42. de Souza, C. B.; Lima, G. P. P.; Borges, C. V.; Dias, L.; Spoto, M. H. F.; Castro, G. R.; Correa, C. R.; Minatel, I. O. Development of a Functional Rice Bran Cookie Rich in Oryzanol. *J. Food Measure. Characterization.* **2019,** *13* (2), 1070–1077.

43. Lu, T.-J.; Chen, H.-N.; Wang, H.-J., Chemical Constituents, Dietary Fiber, and γ-Oryzanol in Six Commercial Varieties of Brown Rice from Taiwan. *Cereal Chemistry* **2011,** *88* (5), 463-466.

44. Jia, M. Y.; Chen, J. J.; Liu, X. Z.; Xie, M. Y.; Nie, S. P.; Chen, Y.; Xie, J. H.; Yu, Q., Structural characteristics and functional properties of soluble dietary fiber from defatted rice bran obtained through *Trichoderma viride* fermentation. *Food Hydrocolloids* **2019,** *94,* 468-474.

45. Sheflin, A. M.; Borresen, E. C.; Kirkwood, J. S.; Boot, C. M.; Whitney, A. K.; Lu, S.; Brown, R. J.; Broeckling, C. D.; Ryan, E. P.; Weir, T. L. Dietary Supplementation with Rice Bran or Navy Bean Alters Gut Bacterial Metabolism in Colorectal Cancer Survivors. *Mol. Nutr. Food Res.* **2017,** *61* (1), 11.

46. Qi, J.; Li, Y.; Masamba, K. G.; Shoemaker, C. F.; Zhong, F.; Majeed, H.; Ma, J. G. The Effect of Chemical Treatment on the In Vitro Hypoglycemic Properties of Rice Bran Insoluble Dietary Fiber. *Food Hydrocolloids* **2016**, *52*, 699–706.

47. Chen, Q. X.; Swist, E.; Kafenzakis, M.; Raju, J.; Brooks, S. P. J.; Scoggan, K. A. Fructooligosaccharides and Wheat Bran Fed at Similar Fermentation Levels Differentially Affect the Expression of Genes Involved in Transport, Signaling, Apoptosis, Cell Proliferation, and Oncogenesis in the Colon Epithelia of Healthy Fischer 344 Rats. *Nutr. Res.* **2019**, *69*, 101–113.

48. Ashwar, B. A.; Gani, A.; Shah, A.; Wani, I. A.; Masoodi, F. A. Preparation, Health Benefits and Applications of Resistant Starch—A Review. *Starch—Stärke* **2016**, *68* (3–4), 287–301.

49. Kahraman, K.; Aktas-Akyildiz, E.; Ozturk, S.; Koksel, H. Effect of Different Resistant Starch Sources and Wheat Bran on Dietary Fibre Content and in Vitro Glycaemic Index Values of Cookies. *J. Cereal Sci.* **2019**, *90*, 6.

50. Malunga, L. N.; Beta, T. Antioxidant Capacity of Arabinoxylan Oligosaccharide Fractions Prepared from Wheat Aleurone Using *Trichoderma viride* or *Neocallimastix patriciarum* xylanase. *Food Chem.* **2015**, *167*, 311–319.

51. Vangsoe, C. T.; Sorensen, J. F.; Knudsen, K. E. B. Aleurone Cells Are the Primary Contributor to Arabinoxylan Oligosaccharide Production from Wheat Bran after Treatment with Cell Wall-Degrading Enzymes. *Int. J. Food Sci. Technol.* **2019**, *54* (10), 2847–2853.

52. Xue, Y. M.; Cui, X. B.; Zhang, Z. H.; Zhou, T.; Gao, R.; Li, Y. X.; Ding, X. X. Effect of β-endoxylanase and α-arabinofuranosidase Enzymatic Hydrolysis on Nutritional and Technological Properties of Wheat Brans. *Food Chem.* **2020**, *302*, 10.

53. Yan, X. G.; Ye, R.; Chen, Y. Blasting Extrusion Processing: The Increase of Soluble Dietary Fiber Content and Extraction of Soluble-Fiber Polysaccharides from Wheat Bran. *Food Chem.* **2015**, *180*, 106–115.

CHAPTER 10

By-Products Derived from Wine Industry: Biological Importance and Its Use

RAMSES M. REYES-REYNA[1,2], ELDA P. SEGURA-CENICEROS[2], RAÚL RODRÍGUEZ-HERERRA[1], ALEJANDRA I. VARGAS-SEGURA[3], JUAN A. ASCACIO-VALDÉS[1], and ADRIANA C. FLORES-GALLEGOS[1*]

[1]Food Research Department, Universidad Autónoma de Coahuila, Boulevard Venustiano Carranza S/N CP 25280 Saltillo, Coahuila, México

[2]Nanobioscience Research Group, School of Chemistry, Universidad Autónoma de Coahuila, Boulevard Venustiano Carranza S/N CP 25280 Saltillo, Coahuila, México

[3]Postgraduate in Advanced Prosthodontics, Dentistry School, Universidad Autónoma de Coahuila, Boulevard Venustiano Carranza S/N CP 25280 Saltillo, Coahuila, México

[*]Corresponding author. E-mail: carolinaflores@uadec.edu.mx

ABSTRACT

The present chapter describes wine by-products and its generation, importance, and future trends to its use. During wine elaboration, high amounts of by-products are generated; actually, approximately 40% are approached. Although wine by-products do not comprise any danger, its inadequate management can cause adverse effects on the environment, being toxic for cultivars, water, and soils because of the presence of tannins. These by-products are a natural source of diverse phytochemicals with important biological activities, these compounds are present majorly in skins and seeds of fruits and are highly valorized due to its effect against degenerative diseases as cancer and cardiovascular illness. Wine by-products have been used in animal feed and for recovering alcohol and tartaric acid; besides, these can be approached due to the viability to generate high-value chemicals such as oils and extracts which can be used in food industry, health, cosmetic, fertilizers, and energy generation.

10.1   INTRODUCTION

Grapes have been present historically since Greek and Roman civilizations and they have been used for the wine industry, fresh consumption, or as dried fruits; today we can find a great variety of grape species within which we can stand out European varieties, North American, and French ones. Grapes also are classified depending on its use as table grapes, wine grapes, dried grapes, with seed or without seed, and grapes for juice production. Grapes are important for its high economic value due to its different presentations; in a minor quantity, we can find processed grape as jelly, juices, vinegar, or grape-seed oil; but the most important use of grape is in wine production, and it is so much that in some countries, grape crops are specifically grown for industrial grape (Torres-Leon et al., 2018; Xia et al., 2013).

Vitis vinifera L. is a species of vine belonging to the family Vitaceae, which lodges a large variety of grapes, among which we can find white and red varieties, which in turn are divided into different strains (Peixoto et al., 2018; Rodríguez Montealegre et al., 2006). The vine, which is the name given to the plant, is a vine-like shrub with a height of approximately 30 cm divided into flowers, fruit, leaves, stem or trunk, and root. The trunk has woody characteristics and loose bark; the roots are deep and multiramified. It also has branches that vary in color from green to reddish-brown and yellow. The flowers have 4–5-mm petals in green and are characterized by multiple clusters. The leaves of the grapevine are slightly oval with three to five lobes, its upper surface is hairless and the lower one is covered with a thin layer of hair (Gruenwald et al., 2000). The fruit, which is named grape, develops in the form of clusters composed of the stalk or scraper, which is the skeleton that gives structure and support to the fruit. It represents between 2% and 7% of the weight of the bunch and the other part that is the grape grain; it is semi-ovoid, its measurements are between 6 and 22 mm long, with color variations ranging from bluish-black, purple, violet and dark red, to green, and yellow, with a flavor that can be sweet and sometimes acidic. The grape skin is the membrane that covers the outside of the fruit and is where the phytochemicals such as tannins, acids, and anthocyanins that play an essential role in the production of the wines are stored since these give it aroma, color, and flavor. Likewise, the skin is covered with a waxy matter called purine, which stores yeasts that will be of high quantity in the fermentation, the skin represents approximately 10% of the grain weight. The seed or nugget is pear-shaped, its oily content results in an unpleasant bitter taste and constitutes 4% of the total weight of the grain. The final part is the pulp of the grain, which is a mass of greenish or white-yellow color, which is the juicy part of the fruit and contains the sugars and aromas of each strain. The grape has nutritional elements such as minerals, vitamins, fibers, carbohydrates, and phytochemicals, of which the most important are polyphenols since they have an important biological activity (Baeza, 2018; Xia et al., 2013). Among the main vitamins that the grape has, are B1, B2, B3, B6, and B9 as well as vitamin A, C, and E; in minerals content they are a source of calcium, phosphorus, sodium, potassium, iron, copper, magnesium, and zinc, on the other hand, the glucose and fructose (SAGARPA, 2017).

Notwithstanding that grapes are a natural source of minerals, vitamins, and fiber their most important compounds are polyphenols, phytochemicals with potent biological activity; these are distributed majorly in grape skins and seeds. Phenols are highly valued due to their antioxidant activity, which has a positive effect against degenerative diseases like certain types of cancer, cardiovascular diseases, and antimicrobial activity (Gruenwald et al., 2000; Lavelli et al., 2017; Peixoto et al., 2018; Xia et al., 2013).

10.2   GRAPE: WORLDWIDE PRODUCTION AND HARVESTING

10.2.1   WORLDWIDE PRODUCTION AND USES

One of the most abundant crops in the world are grapes, the International Organization of Vine and Wine reports a global grape production of 75.1 million tons in 2017. Approximately 50% of grape harvested is used in viniculture industry for fresh table consumption is 37% and at least 9% as dried grapes (Table 10.1) (International Organization of Vine and Wine, 2017).

TABLE 10.1 Continental Grape Use and Production.

Continent	Production per year (t)		
	2012	2013	2014
Africa			
Total	4335.11	4576.95	4654.72
Dried grape	33.83	57.03	47.54
Table grape	2380.06	2455.90	2632.60
Wine	1921.22	2064.02	1974.59
America			
Total	14,542.13	16,115.63	15,150.35
Dried grape	413.48	466.85	418.17
Table grape	3133.44	3380.68	3383.58
Wine	10,995.20	12,268.10	11,348.59
Asia			
Total	21,847.88	23,260.69	23,917.67
Dried grape	964.04	893.73	931.33
Table grape	13,961.59	15,344.00	16,573.32
Wine	6922.26	7022.96	6413.02
Europe			
Total	27,064.66	31,528.83	28,951.13
Dried grape	148.40	135.02	109.97
Table grape	3689.11	3811.89	3993.46
Wine	23,227.15	27,581.93	24,847.70
Oceania			
Total	2077.00	2322.74	2310.60
Dried grape	13.40	7.00	7.00
Table grape	107.70	120.90	133.80
Wine	1955.90	2194.84	2169.80

Globally speaking, China is the biggest producer with 13,083,000 t followed by Italy with a production of 7,169,745 t and the United States of America with 6,679,211 t (Food and Agriculture Organization of the United Nations STAT, 2017). Although China is the number one producer, the largest producer continent is Europe with 35% of world production (Fig. 10.1) (International Organization of Vine and Wine, 2017).

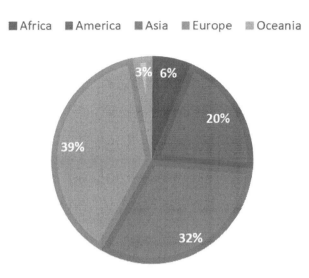

FIGURE 10.1   Worldwide grape production percentage (2014).

Harvesting and quality aspects of grape occur during May–July for table grape and from July to October for industrial grape; grape fruit is used for fresh consumption and industrial grape for juice and wine industries (Mundo and Cada, 2018).

Recently, most countries have grown only varieties of *V. vinifera* species. Three important aspects can influence quality and quantity in wine grapes harvesting: climate, soils, and viticulture practices. Climate conditions limit the grape growing. For a long growing, low environment humidity and sufficient soil moisture are necessary; temperatures are essential for a good development of the vine and are excellent in ripening of grapes: the ideal temperature is between 25 and 30°C, temperatures higher than 38°C stop vine growing, temperatures below 10°C makes the vines remain dormant, and frosts during spring could kill off most of the fruitful shoots. These variations in the climate, topography, and location of the vineyards diversify the wines and their quality.

For grapes, the fertility of soils is not as important as the soil structure, the vines can be cultivated on many soils. The most desirable soils are sandy or gravelly clay loams; it is important to have good drainage and to avoid the alkaline soils. Rootstocks for graft the vine are selected depending on the soil characteristics and to control the vine development. Another aspect to control is the viticulture practices, which include parameters as uniform vine spacing, the vine propagation through grafts or buds, great cultivation, and irrigation, the vegetative parts removing to maintain the vines in a form to control insects and diseases (FAO Investment Centre Division, 2009).

10.3   WINE INDUSTRY: GRAPE TRANSFORMATION INTO WINE

The wine process for red grapes consists of different phases starting immediately after the harvesting; the grapes pass through a destemmer to separate the stems and the leaves from the grains, then the grains are crushed and sent to a fermentation tank. In this place occurs the maceration and alcoholic fermentation because of the yeasts in grape skins, crushed grapes pass approximately 12–15 days in contact with the must, after this process, the wine is separated by gravity for 24 h. The remaining solids are pressed to obtain pressed wine, the solid generated in this stage is called pomace; next step is the malolactic (ML) fermentation of the free-run wine, it consists of the transformation of malic acid into lactic acid trough acid-lactic bacteria, giving a better flavor to the wine. Once ended ML fermentation, racking stage starts, it consists in the separation of the clean wine from the lees, which basically are composed of death yeasts and proteins. After racking, clean wine is transferred to barrels for the aging process, which can last for 12–24 months before bottling (Fig. 10.2).

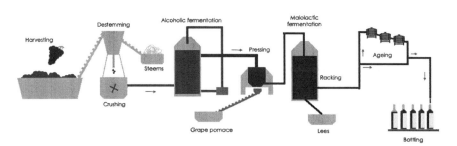

FIGURE 10.2   Wine process and by-products.

For white wines, the process is similar to the red wine: starts after the harvesting, grapes pass through a destemmer to separate stems, and grains are crushed and pressed to separate juice, then the fermentation occurs in the tanks. After fermentation racking is accomplished and wine is transferred to barrels until bottling. This process delivers unfermented grape pomace (GP).

10.4   BY-PRODUCTS DERIVED FROM WINE INDUSTRY: TYPES AND CHARACTERISTICS

Products obtained from grapes, such as wine, juice, jams, raisins, and table grapes, represent an economically important factor: about 80% of the annual world production is used in winemaking, and approximately 20% of the weight of grapes processed remains as waste (Kammerer et al., 2004).

10.4.1   GRAPE POMACE

Grape marc or pomace is the extensive fragment of the solid wastes, corresponds approximately more than 25% of the total weight of the pressed grape, comprises stalks approximately 2%, seeds in 47%, and skins in a 51%. Red GP's primary constituents are organic acids, pigments, lignin, and polyphenols in high quantities, mostly proanthocyanins and tannins. Unfermented pomace contains residues of glucose and fructose (Chikwanha et al., 2018; Muhlack et al., 2017). GP composition can vary depending on factors such as grape variety, climate, and wine processing (must fermentation time). Within the bioactive compounds of the pomace, other components can be found such as dietary fiber (50–75%), polyunsaturated fatty acids in the seeds (10–17%), nonextractable polyphenols (15–30%), and extractable polyphenols (1–9%). Within polyphenols, proanthocyanidins are the most common, as well as catechin, epicatechin, epicatechin gallate, and epigallocatechin which are usually found in the skin and seeds. Particularly in the skin of red grapes are anthocyanins, where the most common are cyanidin, peonidin, petunidin, delphinidin, and malvidin. In lesser amounts exist phenolic acids such as gallic acid, generally present in seeds and stilbenes, such as transresveratrol present in the skin of grapes. As for the seeds, they have important quantities of polyunsaturated fatty acids, such as linoleic acid, the main one being followed by oleic acid and, in a lesser amount, palmitic acid (Lavelli et al., 2017). Among phenolic compounds in pomace, vitamin C exists in high amounts (Sousa et al., 2014).

In spite of the fact that the grape marc does not compromise any danger, if its management is not adequate, it can have adverse effects on the environment, being toxic for crops, contamination of water, and soils due to the presence of tannins, as well as the attraction of flies and disease-causing pests (Chikwanha et al., 2018).

10.4.2   SEEDS

Grape seed contains between 60% and 70% of nondigestible carbohydrates, about 11% of protein, and approximately 19% of oil; it also contains antioxidants like tocopherols and beta carotene which are nonphenolic compounds (Yu and Ahmedna, 2013). Phenolic content in seeds consists of flavan-3-ols mostly, and gallic acid in low concentrations. Polyphenols in seeds also depend on a variety of grapes: for red grapes, as in the case of Shiraz, the main compound in seeds is epicatechin, and for white grapes like Riesling and Viognier are more abundant procyanidin B1. Other factors that can affect the polyphenols concentration in seeds are the berry size and the number of seeds, but also climate, ripeness degree, and variety which could change the composition of grapes and seeds (Rodríguez Montealegre et al., 2006).

10.4.3   GRAPE SKINS

Skins are the main constituent of GP, comprise approximately half of the total matter. Skins' own compounds like polyphenols and triterpenic acids, which can be used in nutritional and pharmaceutical applications. Fermented skins from red grapes contain less pulp and sugars than white grapes because of the ethanolic fermentation. Constituents in skins are inorganic compounds, cellulose, water-soluble compounds, proteins, and tannins (Mendes et al., 2013). The phenolic composition of skins can vary in function of cultivation conditions like weather and soil and the variety of grapes, but the content of anthocyanins and tannins always is going to be higher and gallates content is going to be lower in comparison to other grape parts. The major phenols in grape skins are listed in the following table (Table 10.2) (Pinelo et al., 2006).

TABLE 10.2   Phenols Contained in Grape Skins.

Groups	Phenols
Flavan-3-ols	Catechin, epicatechin, epicatechin gallate, gallocatechin, and epigallocatechin
Anthocyanins	Delphinidin, cyanidin, petunidin, maldivines, and peonidin
Flavonols	Quercetin, kaempferol, myricetin
Hydroxybenzoic acids	Gallic acid and myricetin
Stilbenes	Resveratrol, viniferins
Hydroxycinnamic acids	P-coumaric, caffeic, ferulic, caftaric, coutaric

10.4.4   WINE LEES

Lees or dregs are defined as the semisolid residue obtained by racking after alcoholic fermentation at the bottom of the recipients containing wine, as well as the residue obtained post filtration or centrifugation treatments or during storage (Pérez-Serradilla and Luque de Castro, 2011). Lees are composed of yeasts and bacteria, inorganic matter, phenolic compounds, insoluble carbo-hydrates, lignin, proteins, metals, tartrates, lactic or acetic acids, and tartaric acids; in some cases, there are anthocyanins and phenolic compounds which contribute to the color and flavor of the wine. Lees's content depends on the climate, origins, and variety of grape (Pérez-Bibbins et al., 2015). Lees can be classified as heavy with high amounts of solids and light lees depending on the number of racking process and the particle size (Tao et al., 2014).

10.4.5   VINE SHOOTS

Ninety-three percent of the agroindustry wastes produced per year of viti-culture corresponds to vineyard pruning, which generates vine shots, this process is necessary for equilibration of growing vegetation, it also makes changes in shape, size, quantity, and quality of grapes. This residue is usually burned to prevent phytopathogens proliferation that can cause problems to the environment and in human health (Alfonso et al., 2002). Structure and shape of plantations can influence on the quantity of pruning, vineyards for fresh grape production produce more residues than the vineyards for wine production(Velázquez-Martí et al., 2011). Principal components of vine shots are cellulose, lignin, and hemicellulose in 34%, 27%, and 19%, respectively (Max et al., 2010).

10.4.6   GRAPE STALKS

This by-product corresponds as part of the solid residue during winemaking, stalks are the grape raceme skeleton obtained in the destemming process, represents approximately 6% of the processed raw matter. It consists of tissues composed of lignocellulosic compounds as cellulose, hemicellulose, potassium salts, lignin, and higher amounts of condensed tannins such as procyanidins and prodelphinidins. Variations in compounds on stalks depend directly on the origin, type of grape, season, and climate factors. Due to the high amount of tannins, stalks represent a potential source of antioxidants and for the high hemicellulose content they are a rich source of fermentable sugars. Because of the lignocellulosic content, it is possible to recover glucose, fructose, and xylose. Stalks are characterized by its excellent properties as free radical scavengers because of the high content in phenolic compounds which are more concentrated than other by-products of the wine-making process. Recovering of these compounds signify a reductive impact for the environmental pollution (Cárcel et al., 2010; Garcia-Perez et al., 2010; Ping et al., 2011; Prozil et al., 2012; Prozil et al., 2014; Spigno et al., 2013).

10.5 PHYTOCHEMICALS IN BY-PRODUCTS OF THE WINE INDUSTRY

The most important compounds with the biological activity present in winery by-products are polyphenols within which we find the flavonoids, tannins, and stilbenoids. Grape by-products composition can be variable, for example, red varieties are rich in anthocyanins that are not present in white varieties, and flavan-3-ols such as gallocatechin, procyanidin B1, B2, B4, and C1, catechin and epigallocatechin, are abundant in white varieties. Compounds of grape byproducts comprise principally dietary fiber (50–75% of dry waste) nonextractable polyphenols (15–30%) extractable polyphenols (1–9%) (Saura-Calixto, 1998; Yu and Ahmedna, 2013).

Other phytochemicals in grape by-products are anthocyanins where the most common are cyanidin, peonidin, petunidin, delphinidin, and malvidin; phenolic acids such as gallic acid; stilbenes, such as trans-resveratrol; polyunsaturated fatty acids, such as linoleic acid; oleic acid and, in a lesser amount, palmitic acid (Lavelli et al., 2017). Although stilbenes and phenolic acids are found in less quantity in the pomace, they are concentrated more in red grape varieties than in white ones (Kammerer et al., 2004).

10.6 BIOLOGICAL IMPORTANCE OF COMPOUNDS IN GRAPE BY-PRODUCTS

10.6.1 ANTIOXIDANT ACTIVITY

Grape seeds show the highest amounts of polyphenols as catechin dimers which have a potent antioxidant activity, making it an important product for applications in pharmaceutical, cosmetic, and food industries (Peixoto et al., 2018). The most antioxidative compounds present in wine industry by-products are diphenols, which are effective than simple phenols; due to the increase of polymerization degree, the antioxidant activity also increases (Amico et al., 2008; Soobrattee et al., 2005). The responsible chemical group in phenols for the antioxidant capacity is –OH; the number of these groups and its positions determines the antioxidant capacity (Arora et al., 1998).

10.6.2   CARDIOPROTECTION ACTION

Seed extracts from wine industry waste are a natural source of polyphenols, which are demonstrated to be responsible for the platelet adhesion, and generation of superoxide anion being as effective as resveratrol (Olas et al., 2008). Anthocyanins in grape skins and wine show inhibition for phosphodiesterase-5 activity, which is related to reduce risk of cardiovascular diseases (Dell'Agli et al., 2005). On the other hand, phenols significantly ameliorate the plasma lipid levels and increase the antioxidant capacity of plasma and reduce levels of low-density lipoproteins (Castilla et al., 2006).

10.6.3   ANTIMUTAGENIC AND ANTICANCER PROPERTIES

Grape seed procyanidin extract provides more excellent protection against DNA damage, free radicals, and lipid oxidation. Breast cancer, gastric, and lung adenocarcinoma cells also are attacked by this extract. It also shows protection against skin cancer by the inhibition of the UV radiation-induced oxidative stress (Bagchi et al., 2000).

10.6.4   ANTI-INFLAMMATION ACTIVITIES

Some studies reported that phenolic content in GP inhibits proteases associated with inflammation and cancer (Sartor et al., 2002). Procyanidins in grape seeds demonstrated significant anti-inflammatory effects because of the modulation of adipokine and cytokine gene expression that is related to anti-inflammation (Chacón et al., 2009). Due to the administration of phenolics in patients with hemodialysis treatment, was observed the prevention of inflammation and reduction of plasma monocyte chemoattractant protein 1, a factor involved with cardiovascular diseases (Castilla et al., 2006).

10.6.5   ANTIAGING EFFECTS

Studies with rats demonstrate that polyphenols due to their notable antioxidant activity scavenging free radicals, beneficiates in reverse the course of neuronal and behavioral aging, and prevents organs and tissues from oxidative damage, releases of dopamine from striatal slices and cognitive performance capacity improved notably (Shukitt-Hale et al., 2006). The administration of grape seed extracts shown inhibition of the accumulation of age-related oxidative DNA damages in neural tissue (Balu et al., 2006).

10.6.6   ANTIMICROBIAL EFFECTS

Studies of red wine extract alcohol-free and seed extracts exhibit antimicrobial activity to pathogens such as *Staphylococcus aureus*, *Escherichia coli*, and *Candida albicans* inhibiting the growth of *S. aureus* majorly than *C. albicans* (Papadopoulou et al., 2005; Rotava et al., 2009). Phenols from the different part of grapes indicate a different effect against microbes; however, fermented pomace antimicrobial activity is effective than fresh fruit extracts (Thimothe et al., 2007).

10.7   ACTUAL USE OF GRAPE BY-PRODUCTS

Currently, this residue (in particular, the grape marc) is used for the recovery of tartaric acid or ethanol, generating a new solid residue with high levels of phenolic compounds. Pomace has also been used in animal feed, although this reduces the digestibility and growth of bacteria in the rumen due to the mentioned compounds. Also, its high content of dietary fiber and pectin emphasizes its nutritional value and its possible application as an ingredient in food (Fontana and Antoniolli, 2014).

10.8   FUTURE TRENDS AND PERSPECTIVES

Waste generated by industry (agricultural, food, feed, among others) is the best candidate for bioproducts approach due to the viability to generate high-value substances and platform chemicals. Multiple benefits are promoted to consumers' health. The development of low-cost methods should be explored to retrieve bioactives from winery wastes since several factors determined nutritional composition. The phenolic compounds profile, like tannins, catechins, and phenolic acids, are different in each variety of grapes and therefore, their uses will depend on their compounds. The industry has proposed many theoretical applications for the use of grape bioactives and its consumption is reported to be the itchy scalp, high blood pressure, indigestion, nausea, among others (Patel, 2015). It has been reported that present compounds have good laboratory-level effects as anticancer agents, which makes this residue a promising source to treat these diseases. Also, the National Cancer Institute supports the research involving these materials. Food industry found diverse bioproducts generated. Rivera et al. (2007) mentioned different uses of grape waste as biosurfactant, also glucose in grape can be used as a substrate, pullulan production, and as a food preservative for polyphenolic content, just for mention some from wide applications in food industry (Rivera et al., 2007). Antioxidants and oils extracted from grape waste can be used in the cosmetic industry due to present vitamins for topical application (Dwyer et al., 2014). There are many applications for grape residues, such as in the field of medicine, agriculture, livestock, among others. However, it is still necessary to work on research that generates products that can easily and economically reach consumers, generating beneficial effects in any field of use, mainly in health.

10.9   CONCLUSIONS

The wine industry is highly responsible for environmental impact due to the largest quantities of by-products generated during the process. In last years, the treatment of residues has become a necessity, not as wastes but as a by-product of added value or a primary matter, looking for approaching its content revalorizing wastes as a new source of compounds. It is why many investigations about this by-product have surged for products related to the food industry, health, cosmetics, fertilizers, or energy generation. However, it is necessary to continue and strengthen approaching if these by-products efficiently. Wine residue is partially used for obtaining biocompounds but it is crucial to approach it in its totality due to its content of pectin and fiber into the food industry.

FUNDING SOURCES

This work was supported by the National Council of Science and Technology (CONACYT, México) grant 923202 for Author Ramses M. Reyes-Reyna

Master's studies in the Program in Food Science and Technology offered by the Autonomous University of Coahuila, Mexico.

KEYWORDS

- **antioxidant**
- **biological activity**
- **degenerative diseases**
- **wine by-products**
- **phytochemicals**
- *Vitis vinifera*

REFERENCES

Alfonso, S.; Ysunza, F.; , Beltrán-García, M. J.; Esqueda, M. Biodegradation of Viticulture Wastes by Pleurotus: A Source of Microbial and Human Food and Its Potential Use in Animal Feeding. *J. Agric. Food Chem.* **2002,** *50 (9)*, 2537–2542.

Amico, V.; Chillemi, R.; Mangiafico, S.; Spatafora, C.; Tringali, C. Polyphenol-Enriched Fractions from Sicilian Grape Pomace: HPLC-DAD Analysis and Antioxidant Activity. *Bioresource Technol.* **2008,** *99* (13), 5960–5966. https://doi.org/10.1016/j.biortech.2007.10.037.

Arora, A.; Nair, M. G.; Strasburg, G. M. Structure–Activity Relationships for Antioxidant Activities of a Series of Flavonoids in a Liposomal System. *Free Radical Biol. Med.* **1998,** *24* (9), 1355–1363. https://doi.org/10.1016/S0891-5849(97)00458-9.

Baeza, C. *El libro del vino*; Libsa, E., Ed.; Madrid, 2018.

Bagchi, D.; Bagchi, M.; Stohs, S. J.; Das, D. K.; Ray, S. D.; Kuszynski, C. A.; et al. Free Radicals and Grape Seed Proanthocyanidin Extract: Importance in Human Health and Disease Prevention. *Toxicology* **2000,** *148* (2–3), 187–197. https://doi.org/10.1016/S0300-483X(00)00210-9.

Balu, M.; Sangeetha, P.; Murali, G.; Panneerselvam, C. Modulatory Role of Grape Seed Extract on Age-Related Oxidative DNA Damage in Central Nervous System of Rats. *Brain Res. Bull.* **2006,** *68* (6), 469–473. https://doi.org/10.1016/j.brainresbull.2005.10.007.

Cárcel, J. A.; García-Pérez, J. V.; Mulet, A.; Rodríguez, L.; Riera, E. Ultrasonically Assisted Antioxidant Extraction from Grape Stalks and Olive Leaves. *Phys. Procedia* **2010,** *3* (1), 147–152. https://doi.org/10.1016/j.phpro.2010.01.021.

Castilla, P.; Echarri, R.; Da, A. *Concentrated Red Grape Juice Exerts Antioxidant, Hypolipidemic, and Antiinflammatory Effects in Both*, March 2006; pp 252–262.

Chacón, M. R.; Ceperuelo-Mallafré, V.; Maymó-Masip, E.; Mateo-Sanz, J. M.; Arola, L.; Guitiérrez, C.; et al. Grape-Seed Procyanidins Modulate Inflammation on Human Differentiated Adipocytes in vitro. *Cytokine* **2009,** *47* (2), 137–142. https://doi.org/10.1016/j.cyto.2009.06.001.

Chikwanha, O. C.; Raffrenato, E.; Opara, U. L.; Fawole, O. A.; Setati, M. E.; Muchenje, V.; Mapiye, C. Impact of Dehydration on Retention of Bioactive Profile and Biological Activities of Different Grape (*Vitis vinifera* L.) Pomace Varieties. *Anim. Feed Sci. Technol.* **2018,** *244* (April), 116–127. https://doi.org/10.1016/J.ANIFEEDSCI.2018.08.006.

Dell'Agli, M.; Galli, G. V.; Vrhovsek, U.; Mattivi, F.; Bosisio, E. In Vitro Inhibition of Human cGMP-Specific Phosphodiesterase-5 by Polyphenols from Red Grapes. *J. Agric. Food Chem.* **2005,** *53* (6), 1960–1965. https://doi.org/10.1021/jf048497+

Dwyer, K.; Hosseinian, F.; Rod, M. The Market Potential of Grape Waste Alternatives. *J. Food Res.* **2014,** *3* (2), 91. https://doi.org/10.5539/jfr.v3n2p91.

FAO Investment Centre Division. *Agribusiness Handbook Grapes Wine,* 2009.

Fontana, A. R., & Antoniolli, A. Grape Pomace as a Sustainable Source of Bioactive Compounds: Extraction, Characterization, and Biotechnological Applications of Phenolics. *Agric. Food Chem.* **2013,** *61* (38), 8987–9003. https://doi.org/10.1021/jf402586f.

Food and Agriculture Organization of de United Nations STAT. *Production of Grapes: Top 10 Producers,* 2017. http://www.fao.org/faostat/en/#data/QC/visualize (retrieved January 11, 2019).

Garcia-Perez, J. V.; García-Alvarado, M. A.; Carcel, J. A.; Mulet, A. Extraction Kinetics Modeling of Antioxidants from Grape Stalk (*Vitis vinifera* var. Bobal): Influence of Drying Conditions. *J. Food Eng.* **2010,** *101* (1), 49–58. https://doi.org/10.1016/j.jfoodeng.2010.06.008.

Gruenwald, J.; Brendler, T.; Wyble, C.; Hamid, M.; Nathan, J.; Potter, J. C..; et al. Physician's Desk Reference for Herbal Medicines. *J. Equine Vet. Sci.* **2000,** 19. https://doi.org/10.1016/S0737-0806(99)80323-2.

International Organization of Vine and Wine. *Table and Dried Grapes: World Data Available;* 2017. . http://www.oiv.int/en/oiv-life/table-and-dried-grapes-world-data-available.

Kammerer, D.; Claus, A.; Carle, R.; Schieber, A. Polyphenol Screening of Pomace from Red and White Grape Varieties (*Vitis vinifera L.*) by HPLC-DAD-MS/MS. *J. Agric. Food Chem.* **2004,** *52* (14), 4360–4367. https://doi.org/10.1021/jf049613b.

Lavelli, V.; Kerr, W. L.; García-lomillo, J.; González-sanjosé, M. L. Applications of Recovered Bioactive Compounds in Food Products. *Handbook of Grape Processing By-Products;* Elsevier Inc.: Amsterdam, Netherlands, 2017. https://doi.org/10.1016/B978-0-12-809870-7/00010-7.

Max, B.; Salgado, J. M.; Cortés, S.; Domínguez, J. M. Extraction of Phenolic Acids by Alkaline Hydrolysis from the Solid Residue Obtained after Prehydrolysis of Trimming Vine Shoots. *J. Agric. Food Chem.* **2010,** *58* (3), 1909–1917. https://doi.org/10.1021/jf903441d.

Mendes, J. A. S.; Xavier, A. M. R. B.; Evtuguin, D. V.; Lopes, L. P. C. Integrated Utilization of Grape Skins from White Grape Pomaces. *Industrial Crops Prod.* **2013,** *49*, 286–291. https://doi.org/10.1016/j.indcrop.2013.05.003.

Muhlack, R. A.; Potumarthi, R.; Jeffery, D. W. Sustainable Wineries Through Waste Valorisation: A Review of Grape Marc Utilisation for Value-Added Products. *Waste Manage.* **2017.** https://doi.org/10.1016/j.wasman.2017.11.011.

Mundo, E. L.; Cada, E. S. *Atlas Agroalimentario, 2012–2018,* 2018.

Olas, B., Wachowicz, B.; Tomczak, A.; Erler, J.; Stochmal, A.; Oleszek, W. Comparative Anti-Platelet and Antioxidant Properties of Polyphenol-Rich Extracts from: Berries of Aronia Melanocarpa, Seeds of Grape and Bark of *Yucca schidigera* In Vitro. *Platelets* **2008,** *19* (1), 70–77. https://doi.org/10.1080/09537100701708506.

Papadopoulou, C.; Soulti, K.; Roussis, I. G. Potential Antimicrobial Activity of Red and White Wine Phenolic Extracts against Strains of *Staphylococcus aureus, Escherichia coli,* and *Candida albicans. Food Technol. Biotechnol.* **2005,** *43* (1), 41–46.

Patel, S. Emerging Bioresources with Nutraceutical and Pharmaceutical Prospects. *Emerging Bioresources with Nutraceutical and Pharmaceutical Prospects,* 2015; pp. 1–131. https://doi.org/10.1007/978-3-319-12847-4.

Peixoto, C. M.; Dias, M. I.; Alves, M. J.; Calhelha, R. C.; Barros, L.; Pinho, S. P.; Ferreira, I. C. F. R. Grape Pomace as a Source of Phenolic Compounds and Diverse Bioactive Properties. *Food Chem.* **2018,** *253* (November 2017), 132–138. https://doi.org/10.1016/j.foodchem.2018.01.163.

Pérez-Bibbins, B.; Torrado-Agrasar, A.; Salgado, J. M.; Oliveira, R. P. de S.; Domínguez, J. M. Potential of Lees from Wine, Beer, and Cider Manufacturing as a Source of Economic Nutrients: An Overview. *Waste Manage.* **2015,** *40,* 72–81. https://doi.org/10.1016/j.wasman.2015.03.009.

Pérez-Serradilla, J. A.; Luque de Castro, M. D. Microwave-Assisted Extraction of Phenolic Compounds from Wine Lees and Spray-Drying of the Extract. *Food Chem.* **2011,** *124* (4), 1652–1659. https://doi.org/10.1016/j.foodchem.2010.07.046.

Pinelo, M.; Arnous, A.; Meyer, A. S. Upgrading of Grape Skins: Significance of Plant Cell-Wall Structural Components and Extraction Techniques for Phenol Release. *Trends Food Sci. Technol.* **2006,** *17* (11), 579–590. https://doi.org/10.1016/j.tifs.2006.05.003.

Ping, L.; Brosse, N.; Sannigrahi, P.; Ragauskas, A. Evaluation of Grape Stalks as a Bioresource. *Ind. Crops and Prod.* **2011,** *33* (1), 200–204. https://doi.org/10.1016/j.indcrop.2010.10.009.

Prozil, S. O.; Evtuguin, D. V.; Lopes, L. P. C. Chemical Composition of Grape Stalks of *Vitis vinifera L.* from Red Grape Pomaces. *Ind. Crops Prod.* **2012,** *35* (1), 178–184. https://doi.org/10.1016/j.indcrop.2011.06.035.

Prozil, S. O.; Evtuguin, D. V.; Silva, A. M. S.; Lopes, L. P. C. Structural Characterization of Lignin from Grape Stalks (*Vitis vinifera L.*). *J. Agric. Food Chem.* **2014,** *62* (24), 5420–5428. https://doi.org/10.1021/jf502267s.

Rivera, O. M. P.; Moldes, A. B.; Torrado, A. M.; Domínguez, J. M. Lactic Acid and Biosurfactants Production from Hydrolyzed Distilled Grape Marc. *Process Biochem.* **2007,** *42* (6), 1010–1020. https://doi.org/10.1016/j.procbio.2007.03.011.

Rodríguez Montealegre, R.; Romero Peces, R.; Chacón Vozmediano, J. L.; Martínez Gascueña, J.; García Romero, E. Phenolic Compounds in Skins and Seeds of Ten Grape *Vitis vinifera* Varieties Grown in a Warm Climate. *J. Food Compos. Anal.* **2006,** *19*(6–7), 687–693. https://doi.org/10.1016/j.jfca.2005.05.003.

Rotava, R.; Zanella, I.; Silva, da, P, L., Manfron, M. P., Ceron, C. S.; et al. Antibacterial, Antioxidant and Tanning Activity of Grape By-Product. *Cien. Rural* **2009,** *39* (3), 941–944. Retrieved from http://www.scopus.com/inward/record.url?eid=2-s2.0-67650716052&part nerID=40&md5=71c1bbe70a8173e5b81eaa84a1799fc1.

SAGARPA. Planeación Agrícola Nacional 2017-2030 Uva Mexicana, 2017.

Sartor, L.; Pezzato, E.; Dell'aica, I.; Caniato, R.; Biggin, S.; Garbisa, S. Inhibition of Matrix-Proteases by Polyphenols: Chemical Insights for Anti-Inflammatory and Anti-Invasion Drug Design. *Biochem. Pharmacol.* **2002,** *64* (2), 229–237. https://doi.org/10.1016/S0006-2952(02)01069-9.

Saura-Calixto, F. Antioxidant Dietary Fiber Product: A New Concept and a Potential Food Ingredient. *J. Agric. Food Chem.* **1998,** *46* (10), 4303–4306. https://doi.org/10.1021/jf9803841.

Shukitt-Hale, B.; Carey, A.; Simon, L.; Mark, D. A.; Joseph, J. A. Effects of Concord Grape Juice on Cognitive and Motor Deficits in Aging. *Nutrition* **2006,** *22* (3), 295–302. https://doi.org/10.1016/j.nut.2005.07.016.

Soobrattee, M. A.; Neergheen, V. S.; Luximon-Ramma, A.; Aruoma, O. I.; Bahorun, T. Phenolics as Potential Antioxidant Therapeutic Agents: Mechanism and Actions. *Mutat. Res.—Fundam. Mol. Mech. Mutagen.* **2005,** *579* (1–2), 200–213. https://doi.org/10.1016/j.mrfmmm.2005.03.023.

Sousa, E. C.; Uchôa-Thomaz, A. M. A.; Carioca, J. O. B.; Morais, S. M. de; Lima, A. de; Martins, C. G..; et al. Chemical Composition and Bioactive Compounds of Grape Pomace (*Vitis vinifera L.*), Benitaka Variety, Grown in the Semiarid Region of Northeast Brazil. *Food Sci. Technol. (Campinas)* **2014,** *34* (1), 135–142. https://doi.org/10.1590/S0101-20612014000100020.

Spigno, G.; Maggi, L.; Amendola, D.; Dragoni, M.; De Faveri, D. M. Influence of Cultivar on the Lignocellulosic Fractionation of Grape Stalks. *Ind. Crops Prod.* **2013,** *46*, 283–289. https://doi.org/10.1016/j.indcrop.2013.01.034.

Tao, Y.; Wu, D.; Zhang, Q. A.; Sun, D. W. Ultrasound-Assisted Extraction of Phenolics from Wine Lees: Modeling, Optimization and Stability of Extracts During Storage. *Ultrason. Sonochem.* **2014,** *21* (2), 706–715. https://doi.org/10.1016/j.ultsonch.2013.09.005.

Thimothe, J.; Bonsi, I. A.; Padilla-Zakour, O. I.; Koo, H. Chemical Characterization of Red Wine Grape (*Vitis vinifera* and Vitis interspecific hybrids) and Pomace Phenolic Extracts and Their Biological Activity against *Streptococcus mutans*. *J. Agric. Food Chem.* **2007,** *55* (25), 10200–10207. https://doi.org/10.1021/jf0722405.

Torres-Leon, C.; Ramirez, N.; Londoño, L.; Martinez, G.; Diaz, R.; Navarro, V..; et al. Food Waste and Byproducts: An Opportunity to Minimize Malnutrition and Hunger in Developing Countries. *Front. Sustain. Food Syst.* **2018,** *2* (September), 52. https://doi.org/10.3389/FSUFS.2018.00052.

Velázquez-Martí, B., Fernández-González, E., López-Cortés, I., & Salazar-Hernández, D. M. Quantification of the Residual Biomass Obtained from Pruning of Vineyards in Mediterranean Area. *Biomass Bioenergy* **2011,** *35* (8), 3453–3464. https://doi.org/10.1016/j.biombioe.2011.04.009.

Xia, E.; He, X.; Li, H.; Wu, S.; Li, S.; Deng, G. Biological Activities of Polyphenols from Grapes. *Polyphenols Hum. Health Dis.* **2013,** *1*, 47–58. https://doi.org/10.1016/B978-0-12-398456-2.00005-0.

Yu, J.; Ahmedna, M. Functional Components of Grape Pomace: Their Composition, Biological Properties and Potential Applications. *Int. J. Food Sci. Technol.* **2013,** *48* (2), 221–237. https://doi.org/10.1111/j.1365-2621.2012.03197.x.

CHAPTER 11

Electroanalytical Techniques Applied to Food Analysis Using Nanostructured Sensors

JOSÉ SANDOVAL-CORTÉS[1*], AIDÉ SAÉNZ-GALINDO[2],
J. A. ASCACIO-VALDÉS[3], and C. N. AGUILAR[3]

[1]*Analytical Chemistry Department, School of Chemistry, Autonomous University of Coahuila, Saltillo 25280, Coahuila, México*

[2]*Polymers Department, School of Chemistry, Autonomous University of Coahuila, Saltillo 25280, Coahuila, México*

[3]*Group of Bioprocess and Microbial Biochemistry, School of Chemistry, Autonomous University of Coahuila, Saltillo 25280, Coahuila, México*

Corresponding author. E-mail: josesandoval@uadec.edu.mx

ABSTRACT

This work pretends to leave clearly the basic concepts around of electrochemical nanostructured sensors, given a few examples of some papers published until 2019, going from the definitions of electrochemical sensors to practical applications of developed electrochemical nanostructured sensors fabricated with different nanomaterials also described.

The described nanomaterials are classified in five different kinds, carbon nanotubes, graphene, metal and metal oxide nanoparticles, biomolecules, and molecularly imprinted electrodes. The results described here left no doubts about the convenience when electrochemical techniques are selected, this compared to another typical analytical ways.

11.1   INTRODUCTION

Sensory evaluation is fundamental to introduce new products in the market, but this kind of tests are completely subjective and fully emotional, furthermore the tasters can achieve a great accuracy to determine quality in products like wine or coffee (Cárdenas Mazón et al., 2018). Food quality and food safety requires more precision to determine objective values in order to assure quality and safety parameters, thus the analytical techniques are an excellent option to avoid the subjectivity and emotional face of tester to get precise values for the quality assurance (Benedetti et al., 2004).

Analytical chemistry is the area devoted to qualify and quantify chemical species in a variety of samples, the sample matrix could be simple like drinking water and can be analyzed without any further pretreatment (Lin et al., 2017), or can be complex like meat (Huang et al., 2013) and an extensive pretreatment should be achieved, independently from the nature of the matrix, another way to solve the signals during the analysis could be the selection of the principle of the technique, could be colorimetric, spectrophotometric, chromatographic, fluorometric, titrimetric, nuclear magnetic resonance, or enzymatic (Di Tocco et al., 2018). In general these kinds of methods require the mentioned pretreatment of the sample; in addition expensive equipment is required. On the other hand, electrochemical methods are inexpensive, simple, selective, and reliable (Goda et al., 2019; Li et al., 2019), that is why this chapter is devoted to electrochemical sensors.

One of the most important things in the development of analytical method is the performance of the sensor used in the quantification of the analyte of interest, expressed as selectivity and sensitivity. Better performance can be achieved using nanomaterials in the sensor design. The use of nanomaterials allow us to magnify the signal due to high electrode surface and the decrease of the redox potential due to the facilitated electron transfer process at the electrode surface by nanomaterial inclusion changing the surface physico-chemical properties. This is almost all the times related whit the increment in the sensitivity, achieving lower limits of detection and quantification (Gan et al., 2019; Kumar et al., 2019). As we said, this chapter will be related to electrochemical sensors, especially those conformed by nanomaterials.

11.2   ELECTROCHEMICAL SENSORS

An electrochemical sensor works by electron transfer from the molecule to the electrode (oxidation) or from the electrode to the molecule (reduction).

This kind of electrochemical sensor is called amperometric sensors. This electron transfer occurs at constant electric potential. The electron transfer process is called electrolysis. Electrolysis time vs current or density current graphs corresponds to the technique called chronoamperometry. When the graph shows time of electrolysis vs charge or density charge, the technique is called chronocoulometry. Both could be single or double step, as we said hanging the electrode potential at constant value (Zhou et al., 2017).

If the electrode potential varies vs time, the technique is called voltammetry. Whit this technique, it is possible to determine the redox potential of electrochemical processes. This technique could be single or double cyclic potential scan (Wang et al., 2018). The called potentiometric sensors work measuring changes of the electrode potential that depends on the analyte concentration on the surface electrode. There is no electron transfer between analyte and electrode obtaining potentiometric titration curves (Mittal et al., 2018).

Electrochemical sensors can be classified by the current or potential manipulation as we mention above. However, the response of bare electrode surface could be the same in front of two or more chemical species present in the sample matrix, then the development of current or electrode potential will be the addition of all chemical species. There is one of the most common problems using electrodes as transducers to measure molecules of analytical interest. In order to solve this problem, the chemist do surface modifications attaching different molecules, nanomaterials, or composite materials (Baig et al., 2018; Anu Prathap et al., 2019) in order to obtain a new completely different electrode/solution interface, whit new physicochemical behavior.

Electrochemical sensor can be classified as nanosensors if they include at least one component with nanometric size, or nanobiosensor if the structure includes at least one component with nanometric size and elements like enzymes, antibodies, cell organelles, or proteins (Kumar et al., 2019) between other always coming from a living system.

11.3   NANOSTRUCTURED MATERIALS

There are definitions for nanoscience and nanotechnology, as these words say, the former is the study of phenomena and manipulation of materials at atomic, molecular, and macromolecular scales, and the second one is related to the application (Arfat et al., 2014). To develop applications is important to consider the change between bulk material and the nanomaterial. Their properties will be completely different, nanomaterial has physical and chemical properties suitable form application in many areas such as information technology, materials, medical diagnostics, catalysis, energy, environmental applications (Grassian et al., 2016) and of course food industry (Rodrigues et al., 2017).

Nanometric scale is related determining the enhanced nanomaterials physical or chemical activity, respect to size and surface area, while particle size is decreasing, area is increasing in exponential way (Vaisakh et al., 2016).

Some properties reflect the particle shape and aspect ratio which are clearly different, as example the diffusion of charge transfer between electroactive sites on a modified surface of insulating particles supported onto electrode surface is completely different using spherical, cylinder, and cube-shaped nanoparticles (Eloul et al., 2015).

The characteristic properties exhibited from nanomaterials are directly related to surface charge, also, a specific application could be determined by this property (Espanol et al., 2016), an extreme case is the use of hydroxyapatite adhered to the red blood cells causing aggregation in the unstructured agglutinates caused by surface charge of hydroxyapatite nanoparticles (Han et al., 2012).

Nanoparticles can be formed by different kind of elements, could be from carbon to obtain carbon nanotubes or graphene, or metals to obtain metal or metal oxides nanoparticles, or combination of this kind of material with biomolecules to form nanobiosensor; even more, the constitutional element in a nanosensors could be "nothing" or nanocavities in a polymeric matrix like in a molecularly imprinted sensors.

11.4 CARBON NANOTUBES-BASED SENSORS

Since the carbon nanotubes discover (Iijima, 1991) published in the article "Helical microtubules of graphitic carbon" by Sumio Iijima, the reports about this new and excellent material was incrementing exponentially until 2018 reaching 136,687 articles published.

Searching about the "carbon nanotubes" and "sensor" as keywords throw the SCOPUS, 13,007 articles were published, if the word "electrochemical" is included, the articles published are 4,023, as we can see one third of all articles about sensors with carbon nanotubes were published using some electrochemical technique.

Here, we show some works devoted to the development of nanomaterials prepared with carbon nanotubes to be used to prepare modified electrodes.

The determination of pharmacologically active substances is really important in the food quality assurance (Azzouz et al., 2011). To quantify naproxen, a modified electrode can be fabricated as follows: carbon nanotubes are oxidized by treatment with a mixture of sulfuric and nitric acids and ultrasonication at 40°C, the dispersion obtained was filtrated and washed with cold water and finally dried in a vacuum oven. A new oxidized carbon nanotubes suspension in dimethylformamide was obtained by sonication for 30 min and was used to modify a vitreous carbon electrode surface. The film obtained was useful to oxidize naproxen in a potential 200 mV less than that obtained with the bare electrode. The faradaic current developed by the oxidation process was four times higher using the modified electrode (Montes et al., 2014).

Hydrogen peroxide is used in the food industry as bleacher or to sterilize equipment. Excessive amounts of residual hydrogen peroxide have led to cases of food poisoning (Lu et al., 2011). In order to have an electrode to measure hydrogen peroxide, more complex modification of carbon nanotubes was reported using a polyoxometalate covalently attached to the single carbon nanotubes. Polyoxometalates were reported as catalyzers for the oxidation reactions, took advantage of this property to catalyze the hydrogen peroxide oxidation reaction. The procedure indicates that carboxylated single carbon nanotubes were used to react with the cationic aminopyridinium counter ion alpha-metatungstate $[H_2W_{12}O_{40}]_6$ cluster. Once obtained, the modified single carbon nanotubes, these were attached to the hydroxylated gold electrode surface using 3-aminopropyltrimethoxysilane as binder. Modified gold electrode was evaluated using cyclic voltammetry to observe the hydrogen peroxide oxidation and reduction processes, the synergistic effect due to the presence of carbon nanotubes was clearly observed through the increase of the oxidation and reduction peaks of the polyoxometalate. The sensitivity of hydrogen peroxide detection was increased by a factor of 38.5 in presence of single walled carbon nanotubes (Sahraoui et al., 2019).

11.5 GRAPHENE-BASED SENSORS

Graphene discovery was the last important event in nanotechnology and actually is growing as we can see the numbers reported by SCOPUS. Graphene discovery was published in 2004 in the paper "Electric Field

Effect in Atomically Thin Carbon Films" (Novoselov et al., 2004). Since this year until 2018, 125,753 works were published. The development around the graphene is bigger compared with carbon nanotubes. In the last year, graphene articles 21,564 is double than carbon nanotubes 10,377 and the total papers are almost the same in half time. This data can be observed in Figure 11.1 with data from SCOPUS.

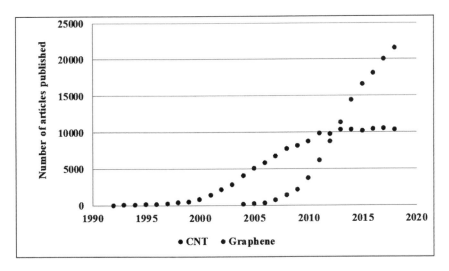

FIGURE 11.1   Number of articles published for carbon nanotubes and graphene since they discover in 1991 and 2004, respectively.

Graphene can be used in two different forms, oxidized graphene or reduced graphene. The group of Alessandra Bonanni presented a study between these different forms to detect quinine, a flavoring agent in tonic beverages. They use graphene oxide, chemically reduced graphene, and thermally reduced graphene. As the results show, the material with the highest current and less overpotential was the chemically reduced graphene and poorest response was for the graphene oxide. This can be explained in relationship with the conjugated system rupture by the oxidation process which includes hydroxy, carbonyl, and carboxyl groups into the graphene structure, diminishing heterogeneous electron-transfer rates on their surface. Then the chemically reduced graphene was employed for the electrochemical detection of quinine in commercial tonic drink samples, showing high sensitivity and selectivity, and therefore representing a valid low-cost alternative to more complicated and time consuming traditional analytical methods (Chng et al., 2017).

Another work presented for the group of Feiyan Yan shows again how the reduced graphene is a better choice to prepare modified electrodes for the electrochemical determinations. They describe the use of nitrogen-doped graphene oxide to determine carbendazim in food samples. The modified electrode was obtained via solvent evaporation from alcoholic nitrogen-doped graphene oxide suspension deposited over glassy carbon electrode; this allows to form a film covering the electrode. The reduction of nitrogen graphene oxide was achieved electrochemically by cyclic voltammetry scanning from 0.0 to -1.0 V at scan rate of 25 mV/s in 0.1 M KNO_3 solution. The oxidized and reduced materials were evaluated and compared with the bare electrode, resulting in the oxidized material better than the bare electrode, but the reduced material exhibited better electrochemical performance than both the bare electrode and nitrogen-doped graphene oxide (Ya et al., 2017).

11.6 METAL AND METAL OXIDE NANOPARTICLES-BASED SENSORS

In order to develop electrochemical methods, a conductive material should be available. The group of Raju Khan demonstrate how the metal nanoparticles can help to achieve this with extraordinary results using the electrode in food analysis. This group used molybdenum disulfide sheets embedded in chitosan, but the molybdenum disulfide semiconducting behavior and nonconducting nature of chitosan do not allow the electrochemical measurement. The use of a conductive element is necessary, gold nanoparticles were the solution, then the electrochemical response was increased. To get a specific sensor for the monosodium glutamate which is a flavoring agent used in food industry, the authors used anti-glutamate antibody attached to the electrode via carbodiimide coupling method. This covalent linkage provokes a reduction in a peak current but is still enough to be used in analytical determinations. This work demonstrates the advantage of metal nanoparticles as electroconductive element in the electrode arrangement (Devi et al., 2019).

The electrochemical platforms for the detection of Sudan I (1-phenylazo-2-hydroxynaphthol), a synthetic lipid soluble azoic pigment used as a food coloring to give red color to the products. The N-doped porous carbon onto glassy carbon electrode was used as electrochemical transducer to detect this colorant but the results were not better than those obtained with the bare electrode, but if copper oxide nanoparticles are introduced the colorant oxidation current in greatly increased. The copper oxide and N-doped porous carbon modified electrode showed high sensitivity and wide linear range in the colorant determination due to the electrocatalytic activity of copper oxide and the electron transfer facilitated by the N-doped porous carbon. This modified electrode was applied in the detection of colorant in chilly sauces and ketchup exhibiting high selectivity, high reproducibility, and stability in the analysis (Ye et al., 2019).

11.7   BIOSENSORS

Biosensors are defined as transducer that contains at least one element related with living cells even more. The complete cells can be introduced into the sensor (Gupta et al., 2019). Most commonly sensors are fabricated from parts of cells or molecules synthesized by cells or part of the biomolecules like aptamers.

Sulfadimethoxine, a commonly used veterinary antibiotic can cause serious problems in the consumers health, the overuse produce accumulation in animal tissues, to achieve the correct measurement of this medication an aptamer-based electrochemical biosensor can be used as reports Zahra Izadi group. They use an electrochemically activated pencil graphite electrode with reduced graphene oxide onto the electrode surface. Over this modified electrode they decorate the electrode with gold nanoparticles formed by electroreduction of $HAuCl_4$ using cyclic voltammetry. Finally, this electrode was immersed into thiolated aptamer solution to be linked covalently to the gold nanoparticles. The prepared electrode was used to measure sulfadi-methoxine in fish, chicken, and beef samples achieving suitable stability, adequate selectivity, and acceptable reproducibility, for sulfadimethoxine determination. The preparation method described can be used to determine any other analyte changing the aptamer sequence (Mohammad-Razdari et al., 2019).

In addition to aptamers, DNA can be used to fabricate electrochemical sensors, the research group of Sani describes the immobilization of amine-terminated single stranded DNA onto silica nanospheres deposited onto electrode with gold nanoparticles previously prepared. The covalent immo-bilization of single-stranded DNA was achieved by the reaction with glutar-aldehyde. This electrode works by methylene blue oxidation. Methylene blue is adsorbed to single strand DNA generating maximum current peak using differential pulse voltammetry technique. This molecular architecture works to detect and quantify carcinogenic molecules like formaldehyde and acrylamide by replacement of methylene blue and the oxidation peak current diminish which is proportional to the carcinogenic compound. The electrode was used to quantify formaldehyde in fish meat and acrylamide in cassava chips. This work presents the study using different DNA sequences resulting carcinogens prefer to bind to guanine base, thus increasing the amounts of guanine bases within the DNA sequence yields better responses in terms of sensitivity, detection limit, linear range, and reproducibility (Sani et al., 2018).

11.8 MOLECULARLY IMPRINTED-BASED SENSORS

As it was said in the introduction, the nanometric recognition element in an electrochemical sensor could be nothing, exactly, just a cavity, an empty space but a space whit the exact form to fit with the analyte of interest. Then to form this target specific cavity, the general procedure is as follows: first, a polymer is produced containing the template or target molecule bound, covalently or noncovalently, to a functional group of the host; second, the template molecule is removed from the polymer host, leaving a target-specific cavity available for rebinding, and finally the molecularly imprinted electrode is exposed to the target-containing sample, and the cavity selectively uptakes the target molecule from a complex sample (Belbruno, 2019).

Estrogens can enter into the human body via food chain and cause adverse health effects. Food industry should be sure that their products should be estrogens free. To achieve this, again, electrochemical methods are a really good option like the electrode fabricated by Hailong Peng group for the 17β-estradiol. This work shows the use of a gold electrode covered with a thick porous layer of gold prepared from a gold silver alloy by dissolution of this last, the membrane has pores with diameter around 20 nm. The polymer used was obtained from 4-Aminothiophenol in a prepolymerization and electropolymerization in the presence of estradiol. This sensor showed excellent selectivity, repeatability, and reusability. Compared with previous reported sensor or HPLC method for detection of 17β-E2, the developed nanowell-based MIP electrochemical sensor is potentially a simple, sensitive, and specific technique for 17β-E2 detection and monitoring for food samples (Wen et al., 2019).

Antioxidants are a group of molecules which can help to avoid oxidation in food. Especially in edible oils can extend the shell life of this kind of products, but an excess of this compound can compromise the quality and safe food products. To measure the quantities of tertiary butylhydro-quinone, a commonly used antioxidant, the Yanhong Bai group developed a modified electrode firstly with graphene oxide dispersion and gold/platinum salts solution to form a film by solvent evaporation, secondly the electroreduction of graphene oxide and gold and platinum cations to obtain electroreduced graphene oxide and gold/platinum nanoparticles to finally end with electropolymerization of o-phenylenediamine and template tertiary butylhydroquinone using cyclic voltammetry scanning method at 50 mV s^{-1} for 10 consecutive cycles between 0 and 1.0 V to fabricate the molecularly imprinted electrode on the surface of electrode. The designed sensor was successfully applied to tertiary butylhydroquinone detection in real edible oil samples, displaying good repeatability and reproducibility. This elec-trode may have great promises for application in antioxidants monitoring in various food samples (Yue et al., 2019).

11.9   CONCLUDING REMARKS

The use of modified electrodes showed in this chapter leave to see that elec-trochemical techniques are cheap, easy, fast, and economic ways to develop methods to be used in food analysis in order to achieve good quality and safe food normativity and quality levels to finally arrive to client satisfaction.

KEYWORDS

- **electrochemical sensors**
- **biosensors**
- **nanotechnology**
- **carbon nanomaterials**
- **metallic nanoparticles**

REFERENCES

Anu Prathap, M. U.; Kaur, B.; Srivastava, R. Electrochemical Sensor Platforms Based on Nanostructured Metal Oxides, and Zeolite-Based Materials. *Chem. Rec.* **2019,** *19*(5), 883–907. https://doi.org/10.1002/tcr.201800068.

Arfat, M. Y.; Zubair, S.; Dar, A. M.; Gatoo, M. A.; Qasim, K.; Naseem, S. Physicochemical Properties of Nanomaterials: Implication in Associated Toxic Manifestations. *Biomed. Res. Int.* **2014**, 1–8. https://doi.org/10.1155/2014/498420.

Azzouz, A.; Jurado-Sánchez, B.; Souhail, B.; Ballesteros, E. Simultaneous Determination of 20 Pharmacologically Active Substances in Cow's Milk, Goat's Milk, and Human Breast Milk by Gas Chromatography–Mass Spectrometry. *J. Agric. Food Chem.* **2011**, *59(*9), 5125–5132. https://doi.org/10.1021/jf200364w.

Baig, N.; Rana, A.; Kawde, A. N. Modified Electrodes for Selective Voltammetric Detection of Biomolecules. *Electroanalysis* **2018**, *30*(11), 2551–2574. https://doi.org/10.1002/elan.201800468.

Belbruno, Joseph J. Molecularly Imprinted Polymers. Review Article. *Chem. Rev.* **2019**, *119*(1), 94–119. https://doi.org/10.1021/acs.chemrev.8b00171.

Benedetti, S.; Pompei, C.; Mannino, S. Comparison of an Electronic Nose with the Sensory Evaluation of Food Products by 'Triangle Test.' *Electroanalysis* **2004**, *16(*21), 1801–1805. https://doi.org/10.1002/elan.200303036.

Cárdenas Mazón, Norma, Carlos Cevallos Hermida, Juan Salazar Yacelga, Efrain Romero Machado, Patricia Gallegos Murillo, Mayra Cáceres Mena. Uso de Pruebas Afectivas, Discriminatorias y Descriptivas de Evaluación Sensorial En El Campo Gastronómico. *Dominio de Las Ciencias* **2018**, *4*(3), 253–263.

Chng, C. E.; Ambrosi, A.; Chua, C. K.; Pumera, M.; Bonanni, A. Chemically Reduced Graphene Oxide for the Assessment of Food Quality: How the Electrochemical Platform Should Be Tailored to the Application. *Chem. Eur. J.* **2017**, *23(*8), 1930–1936. https://doi.org/10.1002/chem.201604746.

Devi, R.; Gogoi, S.; Barua, S.; Dutta, H. S.; Bordoloi, M.; Khan, R. Electrochemical Detection of Monosodium Glutamate in Foodstuffs Based on Au@MoS 2 /Chitosan Modified Glassy Carbon Electrode. *Food Chem.* **2019**, *276*(October 2018), 350–357. https://doi.org/10.1016/j.foodchem.2018.10.024.

Eloul, S.; Compton, R. G. Charge Diffusion on the Surface of Particles with Simple Geometries. *J. Phys. Chem. C* **2015**, *119*(49), 27540–27549. https://doi.org/10.1021/acs.jpcc.5b09455.

Espanol, M.; Mestres, G.; Luxbacher, T.; Dory, J. B.; Ginebra, M. P. Impact of Porosity and Electrolyte Composition on the Surface Charge of Hydroxyapatite Biomaterials. *ACS Appl. Mater. Interfaces* **2016**, *8(*1), 908–917. https://doi.org/10.1021/acsami.5b10404.

Gan, X.; Zhao, H. Understanding Signal Amplification Strategies of Nanostructured Electrochemical Sensors for Environmental Pollutants. *Curr. Opin. Electrochem.* **2019**, *17*, 56–64. https://doi.org/10.1016/j.coelec.2019.04.016.

Goda, E. S.; Gab-Allah, M. A.; Singu, B. S.; Yoon, K. R. Halloysite Nanotubes Based Electrochemical Sensors: A Review. *Microchem. J.* **2019**, *147*(April), 1083–1096. https://doi.org/10.1016/j.microc.2019.04.011.

Grassian, V. H.; Haes, A. J.; Mudunkotuwa, I. A.; Demokritou, P.; Kane, A. B.; Murphy, C. J.; Hutchison, J. E.; Isaacs, J. A.; Jun, Y. S.; Karn, B.; Khondaker, S. I.; Larsen, S. C.; Lau, B. L. T.; Pettibone, J. M.; Sadik, O. A.; Saleh, N. B.; Teague, C. NanoEHS - Defining Fundamental Science Needs: No Easy Feat When the Simple Itself Is Complex. *Environ. Sci. Nano* **2016**, *3*(1), 15–27. https://doi.org/10.1039/c5en00112a.

Gupta, N.; Renugopalakrishnan, V.; Liepmann, D.; Paulmurugan, R.; Malhotra, B. D. Cell-Based Biosensors: Recent Trends, Challenges and Future Perspectives. *Biosens. Bioelectron.* **2019**, *141*(May), 111435. https://doi.org/10.1016/j.bios.2019.111435.

Han, Y.; Wang, X.; Dai, H.; Li, S. Nanosize and Surface Charge Effects of Hydroxyapatite Nanoparticles on Red Blood Cell Suspensions. *ACS Appl. Mater. Interfaces* **2012**, *4*(9), 4616–4622. https://doi.org/10.1021/am300992x.

Huang, X.; Aguilar, Z. P.; Li, H.; Lai, W.; Wei, H.; Xu, H.; Xiong, Y. Fluorescent Ru(phen)32+-Doped Silica Nanoparticles-Based ICTS Sensor for Quantitative Detection of Enrofloxacin Residues in Chicken Meat. *Anal. Chem.* **2013,** *85*(10), 5120–5128.

Iijima, S. Helical Microtubules of Graphitic Carbon. *Nature* **1991,** *354*(6348), 56–58. https://doi.org/10.1038/354056a0.

Mittal, S. K.; Kumar, S.; Kaur, N. Enhanced Performance of CNT-Doped Imine Based Receptors as Fe(III) Sensor Using Potentiometry and Voltammetry. *Electroanalysis* **2018,** 1229–1237. https://doi.org/10.1002/elan.201800100.

Kumar, A.; Purohit, B.; Maurya, P. K.; Pandey, L. M.; Chandra, P. Engineered Nanomaterial Assisted Signal-Amplification Strategies for Enhancing Analytical Performance of Electrochemical Biosensors. *Electroanalysis* **2019,** 1–16. https://doi.org/10.1002/elan.201900216.

Li, P.; Li, X.; Chen, W. Recent Advances in Electrochemical Sensors for the Detection of 2, 4, 6-Trinitrotoluene. *Curr. Opin. Electrochem.* **2019,** *17*, 16–22. https://doi.org/10.1016/j.coelec.2019.04.013.

Lin, W. C.; Li, Z.; Burns, M. A. A Drinking Water Sensor for Lead and Other Heavy Metals. *Anal. Chem.* **2017,** *89*(17), 8748–8756. https://doi.org/10.1021/acs.analchem.7b00843.

Lu, C. P.; Lin, C. T.; Chang, C. M.; Wu, S. H.; Lo, L. C. Nitrophenylboronic Acids as Highly Chemoselective Probes To Detect Hydrogen Peroxide in Foods and Agricultural Products. *J. Agric. Food Chem.* **2011,** *59*(21), 11403–11406. https://doi.org/10.1021/jf202874r.

Mohammad-Razdari, A.; Ghasemi-Varnamkhasti, M.; Izadi, Z.; Rostami, S.; Ensafi, A. A.; Siadat, M.; Losson, E. Detection of Sulfadimethoxine in Meat Samples Using a Novel Electrochemical Biosensor as a Rapid Analysis Method. *J. Food Compos. Anal.* **2019,** *82*(April), 103252. https://doi.org/10.1016/j.jfca.2019.103252.

Montes, R. H. O.; Stefano, J. S.; Richter, E. M.; Munoz Rodrigo, A. A. Exploring Multiwalled Carbon Nanotubes for Naproxen Detection. *Electroanalysis* **2014,** *26*(7), 1449–1453. https://doi.org/10.1002/elan.201400113.

Novoselov, K. S.; Geim, A. K.; Morozov, S. V.; Jiang, D.; Zhang, Y.; Dubonos, S. V.; Grigorieva, I. V.; Firsov, A. A. Electric Field Effect in Atomically Thin Carbon Films Supplementary. *Science* **2004,** *5*(1), 1–12. https://doi.org/10.1126/science.aab1343.

Rodrigues, S. M.; Demokritou, P.; Dokoozlian, N.; Hendren, C. O.; Karn, B.; Mauter, M. S.; Sadik, O. A.; Safarpour, M.; Unrine, J. M.; Viers, J.; Welle, P.; White, J. C.; Wiesner, M. R.; Lowry, G. V. Nanotechnology for Sustainable Food Production: Promising Opportunities and Scientific Challenges. *Environ. Sci. Nano* **2017,** *4*(4), 767–781. https://doi.org/10.1039/c6en00573j.

Sahraoui, Y.; Chaliaa, S.; Maaref, A.; Haddad, A.; Bessueille, F.; Jaffrezic-Renault, N. Synergistic Effect of Polyoxometalate and Single Walled Carbon Nanotubes on Peroxidase-like Mimics and Highly Sensitive Electrochemical Detection of Hydrogen Peroxide. *Electroanalysis* **2019,** 1–8. https://doi.org/10.1002/elan.201900415.

Sani, N. D. M.; Heng, L. Y.; Marugan, R. S. P. M.; Rajab, N. F. Electrochemical DNA Biosensor for Potential Carcinogen Detection in Food Sample. *Food Chem.* **2018,** *269*(July), 503–510. https://doi.org/10.1016/j.foodchem.2018.07.035.

Tocco, A. D.; Robledo, S. N.; Arévalo, F. J.; Osuna, Y.; Sandoval-Cortés, J.; Zon, M. A.; Iliná, A.; Granero, A. M.; Vettorazzi, N. R.; Martínez, J. L.; Segura, E. P.; Fernández, H.

Development of an Electrochemical Biosensor for the Determination of Triglycerides in Serum Samples Based on a Lipase/Magnetite-Chitosan/Copper Oxide Nanoparticles/ Multiwalled Carbon Nanotubes/Pectin Composite. *Talanta* **2018,** *190*(May), 30–37. https:// doi.org/10.1016/j.talanta.2018.07.028.

Vaisakh, S. S.; Mohammed, A. A. P.; Hassanzadeh, M.; Tortorici, J. F.; Metz, R.; Ananthakumar, S. Effect of Nano-Modified SiO2/Al2O3 Mixed-Matrix Micro-Composite Fillers on Thermal, Mechanical, and Tribological Properties of Epoxy Polymers. *Polym. Adv. Technol.* **2016,** *27*(7), 905–914. https://doi.org/10.1002/pat.3747.

Wang, Y.; Cao, W.; Yin, C.; Zhuang, Q.; Ni, Y. Nonenzymatic Amperometric Sensor for Nitrite Detection Based on a Nanocomposite Consisting of Nickel Hydroxide and Reduced Graphene Oxide. *Electroanalysis* **2018,** *30*(12), 2916–2924. https://doi.org/10.1002/ elan.201800627.

Wen, T.; Wang, M.; Luo, M.; Yu, N.; Xiong, H.; Peng, H. A Nanowell-Based Molecularly Imprinted Electrochemical Sensor for Highly Sensitive and Selective Detection of 17β-Estradiol in Food Samples. *Food Chem.* **2019,** *297*(April), 124968. https://doi. org/10.1016/j.foodchem.2019.124968.

Ya, Y.; Jiang, C.; Mo, L.; Li, T.; Xie, L.; He, J.; Tang, L.; Ning, D.; Yan, F. Electrochemical Determination of Carbendazim in Food Samples Using an Electrochemically Reduced Nitrogen-Doped Graphene Oxide-Modified Glassy Carbon Electrode. *Food Analytical Methods* **2017,** *10*(5), 1479–1487. https://doi.org/10.1007/s12161-016-0708-y.

Ye, Q.; Chen, X.; Yang, J.; Wu, D.; Ma, J.; Kong, Y. Fabrication of CuO Nanoparticles-Decorated 3D N-Doped Porous Carbon as Electrochemical Sensing Platform for the Detection of Sudan I. *Food Chem.* **2019,** *287*(August 2018), 375–381. https://doi. org/10.1016/j.foodchem.2019.02.108.

Yue, X.; Luo, X.; Zhou, Z.; Bai, Y. Selective Electrochemical Determination of Tertiary Butylhydroquinone in Edible Oils Based on an In-Situ Assembly Molecularly Imprinted Polymer Sensor. *Food Chem.* **2019,** *289*(October 2018), 84–94. https://doi.org/10.1016/j. foodchem.2019.03.044.

Zhou, K. L.; Wang, H.; Zhang, Y. Z.; Liu, J. B.; Yan, H. Understand the Degradation Mechanism of Electrochromic WO3 Films by Double-Step Chronoamperometry and Chronocoulometry Techniques Combined with in Situ Spectroelectrochemical Study. *Electroanalysis* **2017,** *29*(6), 1573–1585. https://doi.org/10.1002/elan.201700049.

CHAPTER 12

Use of Various Forms of Energy in Food Science and Technology

J. DANIEL GARCÍA-GARCÍA, YESENIA ESTRADA-NIETO,
SANDRA PALACIOS-MICHELENA, ROBERTO ARREDONDO-VALDÉS,
RODOLFO RAMOS-GONZÁLEZ, MÓNICA LIZETH CHÁVEZ-GONZÁLEZ,
MAYELA GOVEA-SALAS, JOSÉ LUIS MARTÍNEZ-HERNÁNDEZ,
ELDA PATRICIA SEGURA-CENICEROS, and ANNA ILYINA*

Nanobioscience group, Chemistry School, Autonomous University of Coahuila, Blvd. V. Carranza e Ing. José Cárdenas Valdés, 25280 Saltillo, Coahuila, Mexico

Corresponding author. E-mail: annailina@uadec.edu.mx

ABSTRACT

The present chapter is focused on the aspects of heat transfer applied in food science and technology. Heat transfer is vital on cooking processes. During food preparation, the energy can be transferred by convection, conduction, and radiation. This chapter describes some forms of heat transfer considering the radiation (microwave and ultrasound) and omics technique as the novel methods for matter transformation during the cooking. Heating is used to destroy microorganisms that can transmit disease or alter food quality, as well as to make food more comfortable to digest. Thus, energy transfer helps to obtain products and systems with new properties, simplifies extraction processes, and matter conversion. Moreover, the advances of nanotechnology provide new nanostructured materials that can be applied as catalysts and supports for the immobilization of enzymes and bioactive compounds. The goal of this review is to analyze the different forms of energy transfer and their application in food processing and extraction of active compounds employed to improve food quality.

12.1   INTRODUCTION

During the transfer, heat flows from an object with a higher temperature to a system with a lower temperature. This process occurs during the thermal treatment of food that is carried out to increase its stability, eliminate the microbial load, cook, or heat food for consumption. Conduction, convection, and radiation are three different heat transfer processes. Traditionally in the food industry, the first two methods are used (conduction and convection), yet not so long ago, the radiation was applied to develop new heat transfer technologies (Fig. 12.1) (Dermikol et al., 2006; Debnath et al., 2012; Gouado et al., 2011).

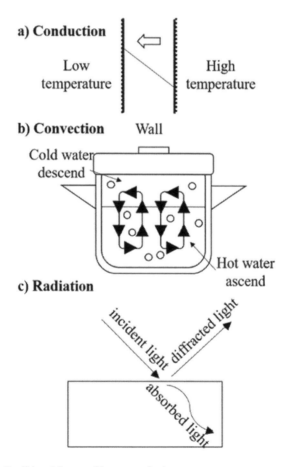

FIGURE 12.1   Traditional forms of heat transferring.

The energy transfer carried out in the presence of temperature gradient is considered as conduction or convection when it takes place in fluids. In both cases, energy flows from higher temperatures to lower. An example is cooking the meat on the grill or a process of boiling in water. Radiation involves the transmission of radiant energy from a transmitter to the receiver system. Absorbed radiation energy leads to increased receiver temperature, for example, cooking in microwaves or drying food in the sun (Santana et al., 2011).

The goal of this review is to analyze the application and mechanisms of different forms of energy transfer used in food processing and extraction of active compounds employed to improve food quality.

12.2   CONDUCTION, CONVECTION, AND RADIATION

Table 12.1 describes some processes used for food preparation and heat transfer phenomena corresponding to each one.

Pasteurization is used for sterilization of liquid foods, for example, milk (Duarte and Cristianini, 2011). There are at least three techniques to carry out the process of pasteurization: (1) the low-temperature method with a long time interval (63–65°C for 30 min); (2) high temperature and short time method (71–75°C for 15 sec); (3) ultrapasteurization method (135–140°C for 2–10 sec). In the process, the heat exchanger systems are used, or sometimes the same container.

Newton's law defines the heat transfer rate (q) of cooling and represents the global effect of convection (Duarte and Cristianini, 2011): $q = hA(T_p - T_\infty)$, where A is heat transfer area; letter h is the coefficient of heat transfer that is defined with the properties of the medium (viscosity, thermal expansion, density) and of the process (pressure, rate, etc.); $(T_p - T_\infty)$ is the temperature gradient; T_p is the temperature of the medium that transfers heat, T_∞ is the temperature of the liquid food.

Commercial sterilization is applied to obtain canned foods that can be preserved without refrigeration for years of storage. During this process, food is kept in a hermetically sealed container at high temperature for a specific time and then cooled.

The Fourier equation defines the heat transfer by conduction in the sterilization process with boundary conditions in correlation with the heat transfer coefficient on the surface of the vessel (Santana et al., 2011): $k \nabla^2 T = \rho Cp (\partial T/\partial t)$ and $k (\partial T/\partial n) = h (T_\infty - T_{surface})$ where k is thermal conductivity, Cp is specific heat, n is number of moles, ρ is density, T is temperature, and t is heating time. When the conduction takes place, the heat flow in conduction is described with the equation:

$$dQ = k\,A\,(-\,\partial t/x)$$ where $(-\,\partial t/\partial x)$ describes a temperature gradient, it has a negative sign, because the highest temperature is outside the container, while the lowest temperature in canned food. Therefore, the instantaneous amount of transfer heat is proportional to area **A** and to the temperature difference ∂t that drives the heat through food thickness ∂x.

TABLE 12.1 Examples of Some Traditional Processes of Heat Transfer Applied in the Food Industry.

Process	Description	Effect on food quality
Pasteurization	Heat treatment of liquid foods applied to decrease the microbial load which mostly applies the convection, and transcendently the conduction.	Nutritional or sensorial changes, which depend on process time and temperature.
Commercial sterilization	Canned food preparation related to the temperature increase and subsequent flushing, which mainly involves the conduction for solid food, the convention for liquid food, and both for liquid food with solid particles.	Nutritional or sensorial changes. Storage time for years.
Cooking in the aqueous medium	Convection, and for large particles— some conduction processes.	Improving digestibility, reducing microbial load and eliminating toxins.
Roasted	Conduction process realized to food processing on a grill	Darkening and loss of some micronutrients.
Frying process	Oil using for heat transfer to the food occurs by conduction and inside food by convection	Loss of moisture and oil gain.
Baking process	Concurrent transfer of heat and matter, which convection, conduction, and radiation.	Loss of moisture at high temperatures.

Cooking in the aqueous medium leads to improve nutritional or sensorial properties of food. During cooking in water, heat transfer is performed by convection on a comparatively low temperature and for showtime due to the water density and condensation temperature. In solid food particles, heat transfer occurs by means of conduction.

During cooking by roast, the heat transfer is carried out by conduction: heat is transmitted from the grill surface (150–180°C) to food. When the food temperature reaches 100°C and more, the water evaporates, and the food darkens due to the conversion of some compounds. The cooking efficiency depends on the degree of heat transfer to the food center. However, for overcooking, the loss of micronutrients may be occurring (Gouado et al., 2011).

Frying is a fast and comfortable heat treatment that often changes food flavors and increases the fat content. In the market, there is a wide variety of fryers that vary in size and capacity. The quality of the products depends on the type of food, process conditions, and kind of oil. The oil serves as a means of heat transfer and influences on the texture and flavor of food (Debnath et al., 2012). A crust forms on the food surface due to dehydration of the surface. The heat transfer rate is a function of the temperature difference of the latter and oil, as well as the coefficient of heat transfer by convection. In the food surface, the heat transfer is performed by convection from oil, and by conduction inside the food. The heat transfer rate is influenced by the liquid oil currents and flow velocity of these currents under temperature 140–180°C that is sufficient for physical and chemical changes in food. The physical-chemical studies that are carried out focus on mass transfer (loss of moisture and oil gain) and energy transfer, which influence the speed of heat transfer.

Baking is a process simultaneous of heat and mass transfers. Table 12.1 shows that the heat transfer is a complex process: (a) heating medium convection; (b) the furnace walls radiation; (c) heating on the inside through conduction. This treatment is characterized by low humidity and high temperature. It is a slow process because the transfer of heat through air convection and radiation of walls is inefficient. Generally, forced convection baking is preferable to the use of oven with induced air circulation. By the ambient air, with flow parallel to the food, the heat is transferred to the food. In the furnace, the heat radiation includes a noncontact method employing electromagnetic heating, for example, infrared radiation. The application of radiation in food science and technology currently has excellent growth and development (Moraga et al., 2011).

12.3   DIELECTRIC HEATING IN FOOD PROCESSING

Induction heating (IH) technology has been in constant technological evolution since the end of 19th and principles of 20th centuries. Michael Faraday discovered the principle of IH when he found the induction of currents by a magnet. However, it was not until James Clerk Maxwell developed the theory of electromagnetism and James Prescott Joule described the heat produced by a current in a conductive material that the fundamental principles of the IH were established (Lucia et al., 2014). On the other hand, dielectric heating (DH) is the process in which the electromagnetic radiation of radio waves or microwaves heats a dielectric (insulating) material. DH is also known as electric heating, radio frequency heating, and high-frequency heating.

IH is a complex noncontact process that combines several phenomena such as electromagnetism, thermal transfer, and metallurgical principles (Rapoport and Pleshivtseva, 2006). IH has many advantages such as high safety, temperature uniformity and control, repeatability, maximum production rate, flexibility and compactness of the heating systems, automation capability, high reliability, energy efficiency, environmental safety, and cost competitiveness, concerning other heating techniques (El-Mashad and Pan, 2016).

IH has gained increasing interest in several industries, such as metalworking, glassware and ceramics processing, semiconductor manufacturing, and chemical synthesis. Likewise, it used in residential, commercial, and industrial kitchens (Rudnev et al., 2017). Besides, IH has been used in applications such as steam production for drying, sterilization, cleaning, and rinsing processes, and even in soil purification (El-Mashad and Pan, 2016). Besides, Pijls et al., (2017) applied IH in biomedical parts to eradicate (sterilizing) bacteria and yeasts from biomaterials with orthopedic applications.

On the other hand, the food industry is one of the most energy-demanding industries for the processing of food products. The most common heating technologies employed are infrared, ohmic heating (OH), and microwave, but IH is a suitable alternative to these conventional technologies. Recently, DH has been used in the processing of food for the reduction of microorganisms and disinfection, cooking, for thawing and tempering (Guo et al., 2019). Table 12.2 shows some important applications of DH in food technology.

TABLE 12.2　Application of Dielectric Heating in Food Technology.

Product	Dielectric heating used to:	References
Milled rice	Disinfestation (against *Corcyra cephalonica*)	Yang et al., 2018
Vegetables	Drying (comparison against microwave technology)	Jiang et al., 2018

TABLE 12.2   *(Continued)*

Product	Dielectric heating used to:	References
Ground beef	Microorganism reduction (against *Escherichia coli*)	Guo et al., 2016; Nagaraj et al., 2016
Broccoli, potato, and salmon	Cooking	Fiore et al., 2013
Tuna (*Thunnusmaccoyii*)	Defrost, thawing	Llave et al., 2015
Apples	Blanching	Manzocco et al., 2008
Shell eggs	Pasteurization (against *Escherichia coli*)	Geveke et al., 2017
Mangoes	Disinfestation (against *Anastrephaludens*)	Sosa-Morales et al., 2009
Pacific saury fish (*Cololabissaira*)	Sterilization (against *Bacillus subtilis*)	Uemura et al., 2017
Kiwi puree	Pasteurization	Lyu et al., 2018
Lean beef meat	Defrosting	Farag et al., 2008, Farag et al., 2009
Frozen shrimp	Tempering	Koray and Miran, 2017
Carrots	Sterilization	Xu et al., 2017
Corn grains	Pasteurization (against *Aspergillus parasiticus*)	Zhen et al., 2017
Dry milk	Sterilization (against *Cronobactersakazakii* and *Salmonella* sp.)	Michael et al., 2014

12.4   OHMIC HEATING, FUNDAMENTS, AND APPLICATIONS

12.4.1   OHMIC HEATING FUNDAMENTS

OH, also known Joule heating, electrical resistance heating, direct electrical resistance heating, electro heating, or electroconductive heating, they are techniques used to generate heat directly to conductive foods through an electric current (Hradecky et al., 2017). OH, is defined as a process where, when applying an electric current, generates heat internally due to its electrical resistance (Alwis and Fryer, 1990). Also, OH is considered different from other electrical methods to generate electrical heating due to in this process the electrodes are put in direct contact with the foods, a difference from the microwave method in which case the electrodes are not used. The applied frequency is less as compared to radio or microwave frequency range, and the waveform is unrestricted, although typically sinusoidal. In this method, the heat inside conductive foods are produced when an electric current is passed through OH; this causes an increase in temperature and allows the generation of energy (Cappato et al., 2017).

In contrast to the most used thermal methods like sterilization, pasteurization, dehydration, and evaporation, by OH the heat generation comes directly from the inside of conductive foods as it is shown in Figure 12.2. Therefore, this technique allows to preserve the sensory characteristics of food due to a shorter heating period, helps to avoid overheating of particles or the surface of the food when it is solid (Kanjanapongkul, 2017). Moreover, OH helps in the production of highly shelf-stable products with proper maintenance of the color and nutritional value of food. It has been used for fruits, vegetables, meat products, milk, flours, and starches (Kaur and Singh, 2016; Lebovka et al., 2006; Sakr and Liu, 2014) because the heat generated inside the food during OH allows rapid and uniform heating (Sakr and Liu, 2014; Loghavi et al., 2009).

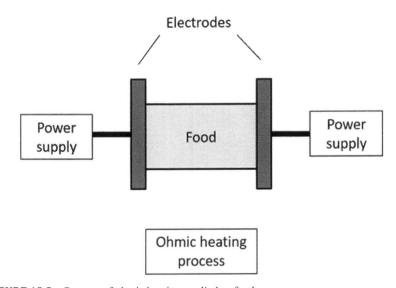

FIGURE 12.2   Process of ohmic heating applied to food.

12.4.2   *FACTORS THAT AFFECT OHMIC HEATING*

However, many factors like electrical conductivity, particle size, field strength, ionic concentration, and electrodes have been found to influence the values of OH rate of feed. The most critical parameter in OH modeling is electrical conductivity (Fryer et al., 1993). When more than one phase is present, these parameters can exert their influence via the effective conductivity of the mixture (σeff) which is the case of particle size and concentration, or they can directly influence the heating rate of the different constituents, which is the case of particle orientation and geometry.

As well, the rate of heating in OH is affected by particle size. Some authors observed that heating rate decreased with the increase in particle size of carrot in an experiment. They pointed out that when the particle size increased, the process temperature had to be increased to achieve the best effect (Zareifard et al., 2003).

In terms of the intensity of the electric field, the greater the electric field intensity, the higher is the electrical conductivity, and the faster is the heating rate. The OH causes damage to some food membranes such as destruction or rupture, for example, in strawberries, and consequently, the free water content increases (De Alwis and Fryer, 1990). Some investigators observed that in order to favor electrical conductivity, the increase in the application of field strength increases the movement of fluids through the capillaries (Halden et al., 1990).

Likewise, the ionic concentration increases in the electrical conductivity during the biological warming of the tissue, causing structural changes that affect the tissue by modifying parameters such as viscosity, gas production (bubbles), firmness of the cell wall (ruptures), or tissue softeners (Sasson and Monselise, 1977). Also, some authors evaluated the effect of ionic concentration on OH of fruits and meat and reported that higher ionic strength has a considerable impact on the heating behavior compared to other techniques (Sarang et al., 2008).

In contrast, the primary heat losses of the periphery of the food material with electrodes depend significantly on the type of cell containing the product and the electrodes, which results in unacceptably high-temperature gradients inside the product. Some researchers report that the cause of heat loss during OH is due to electrodes (Alwis and Fryer, 1990). Other authors reported that the thicker the electrode, lower the rate of temperature increase because it generates a lower electrical resistance by having a larger volume of mass. Moreover, among the most common materials of the electrodes are various metals like stainless steel, titanium, platinized titanium, and aluminum (Zell et al., 2011).

12.4.3   APPLICATIONS OF OHMIC HEATING

12.4.3.1   INACTIVATION OF MICROORGANISMS

In the OH method, the electric field may cause mild nonthermal cellular damage (Pereira et al., 2007). OH is considered lethal to microorganisms due to the low frequency (usually 50–60 Hz) damaging the cell membrane by breaking it and forming pores. What has allowed them to allow the design of experiments and systems that cause the death of pathogenic microorganisms through this technique (Hu et al., 2017; Xiaojing et al., 2017).

The OH has been widely applied to inactivate vegetative cells and the development of some pathogenic microorganisms and the spores produced in various foods such as juices, dairy, and meat, for example, *Bacillus cereus*, *Salmonella sp.*, and *Listeria innocuous*, respectively (Pereira et al., 2007; Zell et al., 2010; Sagong et al., 2011; Kumar et al., 2014; Ryang et al., 2016; Kim and Kang, 2017). Likewise, most of the studies reported with OH compared to conventional procedures indicate that OH generates the best inactivation of microorganisms. Although there are no studies necessary to indicate how this technique affects the biochemical and molecular level of microorganisms (Tian et al., 2018).

12.4.3.2   INACTIVATION OF ENZYMES

The electrical fields applied during OH caused faster inactivation of lipoxy-genase and polyphenoloxidase compared to other conventional techniques (Castro et al., 2004). Similarly, some researchers observed that during OH, one of the mechanisms of microbial inactivation was through an enzymatic process, inactivating pectin esterase, and this occurred by making the resi-dence time shorter because this favors the aromatic compounds involved, they do not degrade as fast as in other conventional methods such as pasteurization or dehydration (Leizerson and Shimoni, 2005).

12.4.3.3   ELECTROPORATION

This process occurs because the electric field of the OH can generate modifications in the cell membrane, altering the dielectric resistance with the membrane beads commonly. In OH, the applied electric field causes electroporation of cell membranes. Several studies report that the amount of lipids present in cell membranes causes a dielectric strength that favors this process (acting as an insulator) present in all the membranes. Variations in the intensity of the electric field can cause a variation in the size of the pores that form in the cell membranes. For this reason, this technique generates greater possibilities of causing the death of a cell, seriously affecting microorganisms. However, these membrane pores can close again in short periods (Marcotte et al., 2008).

12.4.3.4   MEATS

Some authors describe the thermophysical properties of meat products and their behavior during OH (Marcotte et al., 2008). They presented data for thermophysical properties of various meat emulsions. However, these measurements are only in a range from 20 to 80°C. Also, electrical conduction measurements were made on different pieces and cuts of meat using OH cells for analysis (Sarang et al., 2008). Other authors reported that addition of lean-to-fat increased the overall conductivity. In processed meat there are several factors that modify or increase electrical conductivity and reduce OH times, such is the case of some food preservatives (such as sodium chlorides and phosphates), and the amount of endogenous and exogenous fat present in the meats.

12.4.3.5   FRUITS AND VEGETABLE PRODUCTS

During conventional thermal processing of fruits and vegetables, the nutritional quality of most of these products is altered. It was observed that the electrical conductivity decreased in the fruits when modifying the solids content in a mixture of particles, decreasing even more with the bigger particles. The results also suggest that for higher solids (> 20% w/w) and sugar content over 40.0 °Brix, electrical conductivity is too low to use in the conventional OH, and a new design is required. It has already been demonstrated that temperature and thermal processing dramatically affects the appearance, texture, and consistency of various vegetables and fruits. The firmness of processed vegetables (e.g., canned cauliflower, asparagus) can be improved by low temperature, longtime pretreatments (Bartolome and Hoff, 1972; Nimratbir and Singh, 2015; Xiaojing et al., 2018).

12.5   MICROWAVE AND ULTRASOUND

12.5.1   MICROWAVE

Microwave heating is an alternative to conventional heating methods and is an indispensable tool at home. Studies on microwave food technology have been widely addressed in recent decades (Angiolillo et al.; 2015). The first uses of this technology were the fast and effective reheating of food. Heating is not the only application of microwave technology, some other uses such as precooked, tempering frozen foods, pasteurization of food packaging, and drying of pasta products (Tang et al.; 2015; Eke et al.; 2017).

The benefits of microwave heating are not always superior to conventional heating. The benefits depend on the large extent on the type of food matrix being treated and the type of process in general (Atuonwu and Tassou, 2018; Knoerzer et al., 2017). Some of the advantages of microwave heating are shorter heating times, high energy efficiency, precise process control, selective heating, etc. (Ozkoc et al., 2014; Auksornsri et al., 2018). Due to the quick processing times of food through this technology, there is less loss of nutrients compared to exposure to conventional methods in which heating times and temperatures are higher (Auksornsri et al., 2018).

The operation of microwaves is based on the use of electromagnetic frequencies of specific frequencies that generate the heating of the molecules that make up food, the most widely used being those between 300 MHz and 300 GHz (Guo et al., 2017). In conventional methods, heat is applied from the outside to the inside of the food, for microwave heating is from the inside of the food to the outside of the food. One of the essential characteristics of microwave heating is the nonuniformity of heat, generating hot and cold spots (Kumar et al., 2016).

The frequency of the microwaves generates a vibration of the water molecules that are part of the food, causing that it is heated in all its parts simultaneously. The interactions caused by microwave energy and food make heat transferring more effective, and it is why warming times are shorter (Angiolillo et al., 2015). The generated magnetic field promotes that the polarizable molecules or ions present in the food matrix are oriented with respect to the magnetic field (Bhattacharyam and Basak, 2017). During the application of microwaves, hyperthermic conditions occur that affects ionizable and polar molecules (mainly mineral salts and water). Some of the applications in the food area are heating, drying, cooking, reduction, thawing, and inhibition of microbial loading (Chandrasekaran et al., 2013). This last application is because the generated heating conditions interfere in the cell membranes diminishing their physiological activity and survival. Therefore, this method besides being an option for heating is used as a sterilization process. The use of microwave as a method of sterilization or reduction of microbial load has been approved by the Food & Drug Administration (Tang et al., 2015). For this food sterilization tool, it is necessary to have a broad knowledge about the cold points of food (Angiolillo et al., 2015) to ensure correct sterilization.

One of the disadvantages of this method is that its use is not as appreciated in the baked due to obtained textures and flavors, the chemical reactions that give rise to the peculiar organoleptic characteristics of the traditional bakery products (Ozkoc et al., 2014) are not carried out. Some other disadvantages are the development of incomplete taste, loss of moisture, and problems with a firm texture. Studies have been done to improve these problems to try to reduce the combination of heating technologies (Eke et al., 2017; Chizoba-Ekezie et al., 2017). Although the benefits of microwave food processing are known, many people still have doubts about the safety of microwave-processed foods.

12.5.2 ULTRASOUND

High-intensity ultrasound (HIU) is an alternative technology to conventional food processing methods, and it has been adopted as the best technique for reducing processing time and producing quality food (Chemat et al., 2011). HIU is defined as the sound waves that vibrate generating an expansion of the liquid medium and then a contraction of the liquid medium. Depending on the frequencies used in the equipment within the food area, it can be classified as low intensity (>1 MHz) or high intensity (20–100 kHz) ultrasound (Bastarrachea et al., 2017; Awad et al., 2012).

The sound vibrations cause the medium to expand and contract continuously, giving rise to the formation of bubbles that eventually implode. This process is called cavitation (Chemat et al., 2017). The heated environment generated during the cavitation process is generated by chemical and physical changes, these changes include the increase of temperature and pressure, generation of waves that are projected onto the materials with the formation of reactive radicals and reactive oxygen species (Bastarrachea et al., 2017). This process could inactivate enzymes and causes the death of microorganisms (Mamvura et al., 2018; Alzamora et al., 2011). It has been described that it promotes the homogenization and emulsification of ingredients and also enables the extraction of bioactive compounds (Chemat et al., 2011).

Some of the advantages of HIU use are the reduction of processing times, thermal productivity, improvement of the quality attributes of food products, higher yields. Moreover, it is considered as a clean technology due to a decrease in the use of solvents, such as water and lower energy costs. Therefore, this technology is considered as a sustainable alternative for food processing (Chemat et al., 2011; Chemat et al., 2017).

Ultrasound is applied to achieve positive effects (Table 12.3), such as improved mass transfer, food preservation, assistance in heat treatments, and manipulation of food texture and analysis (Bastarrachea et al., 2017). Ultrasound has also been widely used as an extractive method and antimicrobial treatments. The movement generated by the bubbles increases the transfer of mass. During the collision of the bubbles, the cell membranes are broken that allows the diffusion of the used solvent, leading to the extraction process to be faster and more efficient compared to other extraction methods (Ashok Kumar, 2015; Maric et al., 2018).

Ultrasound may be considered a suitable technology for the safe and nutritional quality of food and allows the reduction of energy costs derived from food processing in comparison with conventional technologies.

TABLE 12.3   Uses of Ultrasound Technology in Food Bioprocessing.

Application	References
Improvement of antimicrobial activity of food colorant	Bastarrachea et al., 2017
Estimation of the composition of livestock meat	Awad et al., 2012
Generation of more stable emulsions	Ashok Kumar, 2015
Extraction of bioactive compounds, pigments, and additives	Mark et al., 2018; Ashok Kumar, 2015; Pinela et al.; 2019
Inactivation of enzymes, microorganisms, favors fermentation processes	Ojhav et al.; 2017
Improves homogenization processes	Paniwnyk et al.; 2017
Promotes changes in food coloring	Pingret et al.; 2013
Food cutting and slicing	Pingret et al.; 2013
Determination of carbohydrates content in fruit juices	Kuo et al., 2008
Determination of viscosity	García-Álvarez et al.; 2011
Drying and dehydration of various foods	García-Pérez et al., 2011; Yao, 2016

12.6   SUPERCRITICAL FLUID TECHNOLOGY AND ITS POTENTIAL ON THE FOOD INDUSTRY

Recent global trends show that eco-technologies are necessary to replace conventional ones. Development of alternative "green or eco-friendly" technologies and resulting products required for sustainable processing, energy, and environmental pollution reducing, and finally, to get a healthier society (Nazim-Cifti, 2012). Scientists, industries, and customers have become more interest and consent in the different process at foods where required reduce fats without loss nutritive value or organoleptic properties, with the improving of shelf life and the inhibition of microorganisms. It requires and demands an actual process for nonthermal technologies without the use of preservatives or additives to generate processed food products nutriment (Kumar-Singh et al., 2018).

Supercritical fluid technology (SFT) employs the capability of certain chemicals to become outstanding solvents for specific solutes under a combination of temperature and pressure (Rozzi and Singh, 2002). Fluids become supercritical by increasing pressure and temperature above their critical point. SFT have liquid-like solvent power and gas-like diffusivity. These physical properties would make them ideal clean solvents for processing of natural materials (Nazim-Cifti, 2012), mainly if they applied at relatively high density, that is, at conditions, where molecules are within the range of their mutual interaction potential (Parhi and Suresh, 2012).

A phase diagram in Figure 12.3 for a pure substance demonstrates the temperature and pressure regions where the material stays in a single phase. Such areas bound by curve means the coexistence of two phases (sublimating, deposition, melting, freezing, vaporization, and condensation), and finally, the three curves intersect at the *triple point* (Tp), where all phases coexist in equilibrium (Kumar-Singh et al., 2018). At constant pressure, a phase transition takes place at a transition temperature that is a function by pressure, which involve enthalpy changes. Coexistence curve represents the equilibrium between two phases with a different internal symmetry tended to infinity or eventually intercepted another coexistence curve (Gandhi et al., 2017). In the moment of liquid-gas equilibrium, vapor pressure curve abruptly breaks, recognized like critical point (Cp), that moment requires reaching the critical temperature (Tc), and critical pressure (Pc) above which liquefaction that will not take place through pressure increment, and gas doesn't form with an increase of temperature. This area or region of pressures and temperatures (Pc and Tc) regularly recognized like the *supercritical region* that results in the presence of *supercritical fluid* (Parhi and Suresh, 2012).

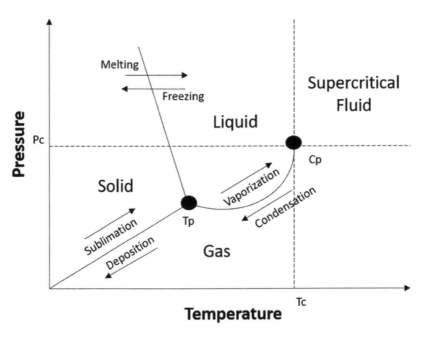

FIGURE 12.3   Solid–liquid–gas phase diagram for pure substance. Tp = triple point. Cp = critical point. Pc= critical pressure. Tc = critical temperature.

SFT extractions present some advantages against traditional extractions methods as higher diffusion coefficient and lower viscosity than liquids. The absence of surface tension allows quick penetration into pores of heterogeneous matrices, which helps enhance extraction efficiencies. The selectivity during extraction may manipulate by fluctuating the conditions of temperature and pressure affecting the solubility of many components. SFT extraction does not leave chemical residue and can use carbon dioxide (CO_2) gas, which can be recycled and used again as part of the unit operation (Rozzi and Singh, 2002).

Carbon dioxide extensively used in SFT by the low toxicity and flammability level characteristics. By another way, CO_2, present some advantages by commercial industry qua is low cost, immediate availability, compatibility solvent properties, and reasonable critical temperature and pressure

(31.2°C and 7.38 MPa) (Cansell et al., 1998). CO_2 products separation can realize achieved by pressure decrease science as products cannot dissolve in CO_2 at normal atmospheric pressure (Nazim-Cifti, 2012). CO_2 technology demonstrated several beneficial and advantageous for processing different food materials. SFT can be applicated on batch extractions of solids technologies, or use for multistage countercurrent fractionation of liquids, then by these ways can be used like by adsorptive and chromatographic fractions. Some potential uses by this technology on food industries developments or procedures systems, actually use on green coffee beans decaffeination, or it is applied on different manufactures of extracts, obtaining several fragrances, aromas, and flavors for a wide variety of plants, herbs, and spices. Other processes in which CO_2 technology can participate consist of fractionation and separate eatable oils components, or contaminant and remnants removal (Brunner, 2005). Currently, due to the food consumption increase, many investigations have been developed to obtain substances by the safe, fast, and low-cost method. In general, the extraction of food compounds is carried out with organic solvents, almost ineffective because they can be toxic, little selective, and severe in use. SFT can get better results because of the ability to extract a higher number of these components with better quality and employing a more efficient process. Table 12.4 shows some applications of the SFT technique in the food area.

TABLE 12.4   SFT of Application in the Food Industry.

Product	A solvent of SFT application	References
Lippiasidoides C., Angelica archangelica L., *Pimpinellaanisum* L., *Piper nigrum* L., *Majoramahortensis* M., *Rosmarinusofficinalis* L., *Stevia rebaudiana* B., *Zingiberofficinale* R., *Rosa aff. Rubiginosa., Capsicumannuum* L., *Theobroma cacao, Paulliniacupana* M.	CO_2 extraction actives like essential oils, antioxidants, gingerols, triglycerides, capsaicinoids, and others	Valle et al., 2005
Pea protein for foam stabilization by aqueous nutriment products	CO_2 extraction	Do Carmo et al., 2016
Eliminating fats from black-eyed pea (*Vigna unguiculata*) and peanut (*Arachishypogaea*) to produce high protein-low fat diet products	CO_2 extraction	Anggrianto et al., 2014
Soybean lecithin—pluronic L64® nanoencapsulated quercetin particles	CO_2 extraction	Lévai et al., 2016

TABLE 12.4 *(Continued)*

Product	A solvent of SFT application	References
PEG particles (Spherical)	Aqueous and CO_2	Markom et al., 2010
Lecithin and β-glucans presence on Resveratrol or another's compounds.	CO_2 extraction	Salgado et al., 2015
Mill waste of olive oil, whole flour, medium oat bran, fine bran, low bran commercial	CO_2 phenolic content	Ty'skiewicz et al., 2018
L. monocytogenes inactivation into dry-cured ham manufactured production.	Where applied several pressures and temperatures of CO_2 at different times. The initial microbial load of *L. monocytogenes* (10^9 CFU/g) reduced below the limit of detection.	Koubaa et al., 2018
Oil ethylated by different algae and fishes	As a technology to fractionate high-valued compounds from lipids with phase silica-pentafluorophenyl (stationary), on mobile phase CO_2	Montañés and Tallon, 2018
Milk and derivate products (buttermilk, butter, cream, or cheese)	Products derived from milk by CO_2 SFT technology present high levels of shelf life combined by satisfactory sensorial properties due to the above current minimal loss of quality characteristics with its microbial inactivation, fat examination, fractionation, content, and solubility, extraction of substances like some cholesterol, vitamins, or flavor's	Kumar-Singh et al., 2018

Use of SFT it is now extended on new industries or process areas like formulation, examination, fraction, and evaluation for several specific treatments chemical reactions. Costs of SFT extraction (SFTE) processes are competitive. In some instances, SFTE is an exclusive way to meet product conditions (Brunner, 2005). SFT technologies offer several advantages against traditional organic solvent technologies, such as eco-friendly, fast, and simple product separation or fraction. There are some manufacturing elements for the separate components from liquid combinations using subcritical fluid or SFT. The central benefits of SCF use for separation natural products consist of solvent-free conditions, no by-products presence, and low temperatures throughout the separation method.

12.7   OTHER POTENTIAL TECHNOLOGIES IN FOOD PROCESSING

12.7.1   NANOPARTICLES FOR EXTRACTION OF BIOACTIVE COMPOUNDS

Nowadays, nanoparticles (NPs) are not entirely accepted in human use. The creation of NPs hazards regulation to human health has become a priority for governments and researches. However, NPs represent a potential technology to be used in food or as excipients in pharmacology. It is the reason why the synthesis of nanomaterials like NPs has taken significant importance in science. Nanomaterials can provide solutions to many technological and environmental challenges in the field of solar energy conversion, medicine, and wastewater treatment (Singh and Tandon, 2014). In the process of global efforts to reduce hazardous waste, there is need to develop synthesis routes which are economical, cost-effective, nontoxic, and productive. NPs are particles sized between 1 and 100 nm. NPs may be synthesized by several methods divided in two significant ways like (i) top-down technologies and (ii) bottom-up technologies (Singh and Tandon, 2014; Chaturvedi et al., 2012). Various sophisticated instruments have been used to characterize the nanomaterials to find out actual size, shape, surface structure, valence, chemical composition, electron bandgap, bonding environment, light emission, adsorption, scattering, and diffraction properties.

The use of NPs to catalyze reactions has undergone explosive growth, inhomogeneous and heterogeneous catalysis. NPs have a high surface-to-volume ratio compared to bulk materials. They are attractive to use as catalysts. Catalysts daily accelerate and boost thousands of different chemical reactions, and thereby form the basis for the multibillion dollar worldwide chemical industry and environmental protection technologies. Research in nanotechnology and nanoscience is expected to have a significant impact on the development of new catalysts. The detailed understanding of the chemistry of nanostructures and the ability to control materials on the nanometer scale could ensure rational and cost-efficient development of new and more capable catalysts for chemical processes.

The magnetic nanoparticles (MNPs) have the advantage of being manipulated by the application of an external magnetic field. They have physical and chemical properties that turn out to be attractive for their use in different areas, such as catalysis, biotechnology, biomedicine, electronics, etc. (Zhan et al., 2018). However, the main disadvantage is their chemical instability, which is why they are usually coated with stable materials to maintain their original properties and appropriately stabilizing them. This chemical instability is because NPs have a very high surface to volume ratio. When the surface-volume is small, the NP could have a higher specific surface area, and the total volume of atoms could be minimal; therefore, the behavior of these could be unstable. For this reason, MNP are coated with inorganic materials such as silica, presenting higher stability, leading to the formation of core-shell magnetic NPs with a wide range of applications (Chaturvedi et al., 2012) in areas such as: biology, mechanics, optics, magnetism, chemistry, catalysis (photocatalysis), solar cells, and microelectronics.

The use and application of magnetic NPs originated from the immobilization and magnetic manipulation of enzymes linked to insoluble supports (Zhan et al., 2018). Robinson used magnetic particles of silica iron oxide and an iron-cellulose oxide composition for the respective immobilization of α-chymotrypsin and β-galactosidase (Lu et al., 2007). A decade later, it was discovered that the interaction between the DNA and the surface of silica microspheres created ionic bonds that kept both structures together. It led to the development of purification methods which take advantage of this behavior. Thus, in conjunction with the development of nanotechnology and other disciplines in the field of science and engineering, the techniques of magnetic NP synthesis with optimal properties have been investigated and improved for their application in the magnetic extraction of DNA (Mélinon et al., 2014).

There are many types of magnetic particles such as magnetite, cobalt ferrite, zinc ferrite, or cobalt–zinc ferrite, among others, which have a super paramagnetic behavior when the NP sizes are less than 10 nm. Other applications that present the nanostructured systems type core-shell with magnetic properties are in the transport and controlled release of drugs, and in nuclear magnetic resonance imaging (Gawande et al., 2015).

Therefore, magnetic NPs have been widely used in extraction techniques, in particular, as magnetic sorbents for extraction in solid dispersive phase (DSPE) or dispersive solid-phase microextraction (DSPME). The availability of these NPs has allowed the development of microextraction in the magnetic dispersive solid phase (DMSPME) which involves the use of a magnetic nanosorbent uniformly dispersed in the sample. The analytes are adsorbed on the nanosorbent during agitation and then are recovered from the solution using a magnetic field. Once the nanosorbent is recovered with the analytes, the dissolution is eliminated. Subsequently, the magnetic phase is dispersed in the elution solvent. After the elution of the analytes from the nanosorbent, the magnet is used again to separate the nanosorbent from the solution.

In some cases, the extraction or elution stages are assisted by ultrasound or vortex shaking, among others. The importance of this technique is mainly due to its high preconcentration factor, simplicity, a high surface area ratio of the nanosorbent, the extraction time decrease without centrifugation, or filtration step. Solid-phase extraction (SPE) with NPs is a rapidly growing subfield (Chen et al., 2010). In the context of metabolomics, it is an emerging approach to selectively separate and enriches metabolites from complex mixtures. The underlying principle of SPE is that the adsorption of targeted analytes on a solid surface (stationary phase) from a solution (mobile phase) can be reversed through a second mobile phase based on physical-chemical properties of the metabolites adsorbed to the solid surface and the characteristics of the mobile phase. Compared to organic solvent extraction, selective metabolite extraction decreases the variety of extracted compounds and improves their quality through repeatability and reduced matrix effects (Zhan et al., 2018). For example, MNPs can be used as SPE sorbent, and their surface can also be modified to isolate and concentrate specific trace analytes of interest from complex matrices. Instead of centrifugation or filtration, an external magnetic field is employed to separate the sorbent-analyte complexes from the solvent before the analytes are released and concentrated.

Today, several commercial or synthetic nanomaterials including functionalized silica, carbon nanotubes, graphene oxides, zeolites, and nanofibers have been applied as sorbents in the DSPME. These nanomaterials can be used for the enrichment and separation of several analytes from different matrices. The high surface area value of these nanomaterials increases the kinetics of extraction, in addition to a wide variety in the applicability. The modification of these nanomaterials with MNPs makes it easier to handle them and avoid the subsequent use of centrifugation or filtration (Campbell et al., 2011).

12.7.2   NANOENCAPSULATION

Nanoencapsulation has been used widely in food, cosmetic, pharmaceutical, and textile industries. Nanoencapsulation is defined as the entrapping of bioactive compounds in nanometric particles. This process involves packing of solid, liquid, or gas material in second material called carrier. Nanoencapsulates can be obtained by physical, chemical, and physicochemical methods. According to the carrier material and the applied method, nano-encapsulated samples could be mononuclear, polynuclear, or matrix, as it is shown in Figure 12.4, and size could range from 1 to 1000 nm (Cano-sarabia and Maspoch, 2015).

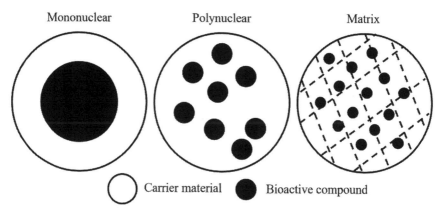

FIGURE 12.4   Type of structures for nanoencapsulation of bioactive compounds.

12.7.2.1   SPRAY DRYING

Spray drying is one of the most used methods due to it is simple, fast, cheap, yielding powder. High amounts of material can be dried. There is an excellent variety of carrier material and bioactive compound to use (Murugesan and Valérie, 2012). Through spray drying is possible to get powders with 90% of the moisture to reach them more stables for a long time. This method involves three main processes: preparation, homogenization, and aspersion of obtained dispersion. However, the high temperatures used in the dryer can negatively effect on bioactive compounds (Mensink et al., 2017). Crystalline powders have an ordered structural location, which allows them to be stable. To disrupt this arrangement, external energy must be applied, for example, during the dissolution of crystalline powders in the solvents, like water. Spray drying used temperatures from 50 to 220°C. On the other hand, it is well known that solubility increase at high temperatures. For this reason, solubility becomes an essential factor to consider in feed liquid designing to apply for spray drying. Once carrier material and bioactive compounds are dissolved in solution, water removal is the next step. The liquid is feed by a nozzle and is atomized, droplets are saturated with water, and as long as they are traveling along with the drying chamber, the drying process is carried out. A successful procedure is achieved when water removal takes place rapid enough, and, in this way, stickiness powder is avoided. Stickiness is because the dry temperature is higher than the glass transition temperature (T_g), which lead to a thermoplastic behavior. The temperature on the droplet surface should not be above T_g to avoid sticky powders (Solans and García-Celma, 2017). Therefore, carrier material must be selected carefully, considering these parameters.

12.7.2.2 MICROEMULSION AND NANOEMULSION

Microemulsions and nanoemulsions have gained importance in different industries. They are thermodynamically stable isotropic colloidal solutions of two immiscible liquids (i.e., water and oil) obtained with surfactant molecules with appropriate hydrophilic–lipophilic properties. These emulsions are characterized by a transparent/translucent aspect, high interfacial area, and low viscosity (Solans and García-Celma, 2017). It was demonstrated that the cosurfactant slowly added in enough quantity into a suspension of water/oil provides a stable emulsion (Horsak and Slama, 1987). One of the main advantages is the size of the obtained particles, ranges from 20 to 500 nm is achieved, and agglomeration is avoided {Formatting Citation}. Microemulsions are generally classified as microemulsion (organic phase/water) and inverse microemulsion (water/organic phase) based on the ratio of their components. Inverse microemulsion or reverse microemulsion is formed when water is dispersed in oil. It contains three main components: water, nonpolar organic phase, and surfactant, which must be used in a suitable proportion according to the triangular phase diagram (Horsak and Slama, 1987). Triangular phase diagram, known as ternary phase diagrams, graphically depicts the proportion of component in an equilateral triangle. The most common use is to show the composition of systems formed by three elements. In a ternary plot, the sum of the three variables a, b, and c must be equal to a constant (K), which generally is 100%. Therefore, it is possible to calculate missing values having two of them. This graph allows us to determine the concentration ranges in which microemulsion is stable, the minimum and maximum proportion of each component. Therefore, values out of this range are useful for recovering products synthesized (Liu et al., 2004).

12.7.2.3   NANOPRECIPITATION

Nanoprecipitation method is carried out by the formation of insoluble phase. The organic solvent is evaporated to concentrate on nano-precipitates (Dimer et al., 2013). Lyophilization is used to eliminate the aqueous phase. One of the main challenges is the agglomeration. Nanoprecipitation allows the using of surfactant, which helps to avoid particle agglomeration and to obtain smaller particle sizes (Martínez et al., 2017). Nanoprecipitation is performed in four steps: supersaturation, nucleation, growth, and coagulation. Supersaturation rate can affect NPs size: the more level of supersaturation, the smaller particle size is obtained. Stability is achieved in the nucleation step, which is finished when the solute concentration decreases, due to nuclei growth, coagulation, or condensation. Condensation is the addition of single molecules to the particle surface, while coagulation is when particles adhere to each other by Van Der Walls or hydrophobic interactions (Dimer et al., 2013; Hornig et al., 2009).

12.8   CONCLUSION

Heat transfer is the method to make the food safe to eat and more comfort-able to digest. Currently, the novel methods for matter transformation for the cooking are developed and used: microwave, ultrasound, omics, and DH. SFT and nanotechnology approaches are potential tools to obtain products and systems with new properties, simplify extraction processes, and matter conversion. Additional efforts must be applied to broaden the spectrum of methods that use new technologies and establish adequate conditions for their implementation.

KEYWORDS

- **physical-chemical properties**
- **food processing**
- **heat**
- **nanotechnology**
- **food quality**

REFERENCES

Alwis, A. A. P.; Fryer, P. J. The Use of Network Theory in the Finite-Element Analysis of Coupled Thermo-Electric Fields. *Commun. Appl. Num. Meth.* **1990,** *6,* 61–66.

Alzamora, S. M.; Guerrero, S.; Schenk, M.; Raffellini, S.; López-Malo, A. Inactivation of Microorganism. In *Ultrasound Technologies for Food and Bioprocessing*; Feng, H., Hao, Barbosa-Cánovas, G., Weiss, J., Eds.; Springer: New York, 2011; pp 321–344.

Anggrianto, K.; Salea, R.; Veriansyah, B.; Tjandrawinata, R. R. Application of Supercritical Fluid Extraction on Food Processing: Black-Eyed Pea (*Vigna unguiculata*) and Peanut (*Arachis hypogaea*). *Procedia Chem.* **2014,** *9,* 265–272.

Angiolillo, L.; Del Nobile, M. A.; Conte A. The Extraction of Bioactive Compounds from Food Residues using Microwaves. *Curr. Opin. Food Sci.* **2015,** *5,* 93–98.

Ashokkumar, M. Applications of Ultrasound in Food and Bioprocessing. *Ultrason. Sonochem.* **2015,** *25,* 17–23.

Atuonwu, J. C.; Tassou, S. A. Quality Assurance in Microwave Food Processing and the Enabling Potentials of Solid-State Power Generators: A Review. *J. Food Eng.* **2018,** *234,* 1–15.

Auksornsri, T.; Tang, J.; Tang, Z.; Lin, H.; Songsermpong S. Dielectric Properties of Rice Model Food Systems Relevant to Microwave Sterilization Process. *Innov. Food Sci. Emer. Technol.* **2018,** *45,* 98–105.

Awad, T. S.; Moharram, H. A.; Shaltout, O. E.; Asker, D.; Youssef, M. M. Application of Ultrasound in Analysis, Processing and Quality Control of Food: A Review. *Food Res. Intern.* **2012,** *48,* 410–427.

Bartolome, L. G.; Hoff, J. H. Firming of Potatoes: Biochemical Effects of Pre-Heating. *J. Agric. Food Chem.* **1972,** *20,* 266–270.

Bastarrachea, L. J.; Walsh, M.; Wrenn, S. P.; Tikekar, R. V. Enhanced Antimicrobial Effect of Ultrasound by the Food Colorant Erythrosin B. *Food Res. Int.* **2017,** *100,* 344–351.

Bhattacharyam, M.; Basak, T. A Comprehensive Analysis on the Effect of Shape on the Microwave Heating Dyinamics of Food Materials. *Innov. Food Sci. Emegr. Technol.* **2017,** *39,* 247–266.

Brunner, G. Supercritical Fluids: Technology and Application to Food Processing. *J. Food Eng.* **2005,** *67,* 21–33.

Campbell, J. L.; Arora, J.; Cowell, S. F.; Garg, A.; Eu, P.; Bhargava, S. K.; Bansal, V. Quasi-Cubic Magnetite/Silica Core-Shell Nanoparticles as Enhanced MRI Contrast Agents for Cancer Imaging. *PLOS ONE* **2011,** *6*(7).

Cano-sarabia, M.; Maspoch, D. Nanoencapsulation. Encyclopedia of Nanotechnology **2015**, *50*, 1–16. http://choicereviews.org/review/10.5860/CHOICE.50-3014.

Cansell, F.; Rey, S.; Beslin, P. Thermodynamic Aspects of Supercritical Fluids Processing: Applications to Polymers and Wastes Treatment. *Rev. Inst. Fran. Pétr.* **1998**, *53*(1), 71–98.

Cappato, L. P.; Ferreira, J. T.; Guimaraes, J. B.; Portela, A. L. R.; Costa, M. Q.; Freitas, R. L.; Cunha, C. A. F.; Oliveira, G. D.; Mercali, L. D. F.; Marzack, A. G. Ohmic Heating in Dairy Processing: Relevant Aspects for Safety and Quality. *Trends Food Sci. Technol.* **2017**, *62*, 104–112.

Castro, I.; Teixeira, J. A.; Salengke, S.; Sastry, S. K; Vicente, A. A. Ohmic Heating of Strawberry Products: Electrical Conductivity Measurements and Ascorbic Acid Degradation Kinetics. *Innov. Food Sci. Emerg. Technol.* **2004**, *5*, 27–36.

Chandrasekaran, S.; Ramantathan, S.; Basak, T. Microwave Food Processing-A Review. *Food Res. Int.* **2013**, *52*, 243–261.

Chaturvedi, S.; Dave, P. N.; Shah, N. K. Applications of Nano-Catalyst in New Era. *J. Saudi Chem. Soc.* **2012**, *16*, 307–325.

Chemat, F.; Huma, Z.; Khanm, M. K. Applications of Ultrasound in Food Technology: Processing; Preservation and Extraction. *Ultrason. Sonochem.* **2011**, *18*, 813–835.

Chemat, F.; Rombaut, N.; Sicaire, A. G.; Meullemiestre, A.; Fabiano-Tixier, A. S.; Abert-Vian, M. Ultrasound Assisted Extraction of Food and Natural Products. Mechanisms; Techniques; Combinations; Protocols and Applications. A Review. *Ultrason.Sonochem.* **2017**, *34*, 540–560.

Chen, F.; Shi, R.; Xue, Y.; Chen, L.; Wan, Q. H. Templated Synthesis of Monodisperse Mesoporous Maghemite/Silica Microspheres for Magnetic Separation of Genomic DNA. *J. Magnet. Magn. Mater.* **2010**, *322*(16), 2439–2445.

ChizobaEkezie, F. G.; Sun, D. W.; Han, Z.; Cheng, J. H. Microwave-Assisted Food Processing Technologies for Enhancing Product Quality and Process Efficiency: A Review of Recent Developments. *Trends Food Sci. Technol.* **2017**, *67*, 58–69.

De Alwis, A. A. P.; Fryer, P. J. A Finite-Element Analysis of Heat Generation and Transfer During Ohmic Heating of Food. *Chem. Eng. Sci.* **1990**, *45*(6), 1547–1559.

Debnath, S.; Rastogi, N. K.; Gopala, A. G.; Lokesh, B. R. Effect of Frying Cycles on Physical, Chemical and Heat Transfer Quality of Rice Bran Oil During Deep-Fat Frying of Poori: An Indian Traditional Fried Food. *Food Bioprod. Proces.* **2012**, *90*, 249–256.

Dermikol, E.; Erdogdu, F.; Koray, T. A Numerical Approach with Variable Temperature Boundary Conditions to Determine the Effective Heat Transfer Coefficient Values During Baking Cookies. *J. Food Proc. Eng.* **2006**, *29*, 478–497.

Dimer, F. A.; Pohlmann, A. R.; Guterres, S. S. Characterization of Rheology and Release Profiles of Olanzapine-Loaded Lipid-Core Nanocapsules in Thermosensitive Hydrogel. *J. Nanosci. Nanotech.* **2013**, *13*(12), 8144–8153.

Do Carmo, C. S.; Nunes, A. N.; Silva, I.; Maia, C.; Poejo, J.; Ferreira-Dias, S.; Duarte, C. M. M. Formulation of Pea Protein for Increased Satiety and Improved Foaming Properties. *RSC Adv.* **2016**, *6*, 6048–6057.

Duarte, P. E.; Cristianini, M. Determining the Convective Heat Transfer Coefficient (h) in Thermal Process of Foods. *Int. J. Food Eng.* **2011**, *7*(4), 1–22.

Eke, B. C.; Jibiri, N. N.; Bede, E. N.; Anusionwu, B. C.; Orji, C. E. Effect of Ingestión of Microwaved Foods on Serum Anti-Oxidant Enzymes and Vitamins of Albino Rats. *J. Rad. Res. Appl. Sci.* **2017**, *10*, 148–151.

Farag, K. W.; Duggan, E.; Morgan, D. J.; Cronin, D. A.; Lyng, J. G. A Comparison of Conventional and Radio Frequency Defrosting of Lean Beef Meats: Effects on Water Binding Characteristics. *Meat Sci.* **2009**, *83*(2), 278–284.

Farag, K. W.; Lyng, J. G.; Morgan, D. J.; Cronin, D. A. A Comparison of Conventional and Radio Frequency Tempering of Beef Meats: Effects on Product Temperature Distribution. *Meat Sci.* **2008**, *80*(2), 488–495.

Fiore, A.; Di Monaco R.; Cavella, S.; Visconti, A.; Karneili, O.; Bernhardt, A.; Fogliano, V. Chemical Profile and Sensory Properties of Different Foods Cooked by a New Radiofrequency Oven. *Food Chem.* **2013**, *139*(1–4), 515–520.

Fryer, P. J.; de Alwis, A. A. P.; Koury, E.; Stapley, A. G. F.; Zhang, L. Ohmic Processing of Solid–Liquid Mixtures: Heat Generation and Convection Effects. *J. Food Eng.* **1993**, *18*, 101–125.

Gandhi, K.; Arora, S.; Kumar, A. Industrial Applications of Supercritical Fluid Extraction: A Review. *Int. J. Chem. Stud.* **2017**, *5*(3), 336–340.

García-Álvarez, J.; Salazar, J.; Rosell, C. M. Ultrasonic Study of Wheat Flour Properties. *Ultrasonics* **2011**, *51*, 223–228.

García-Pérez, J. V.; Ozuna, C.; Ortuño, C.; Cárcel, J. A.; Mulet, A. Modeling Ultrasonically Assisted Convective Drying of Eggplant. *Dry. Technol.* **2011**, *29*, 1499–1509.

Gawande, M. B.; Goswami, A.; Asefa, T.; Guo, H.; Biradar, A. V.; Peng, D. L.; Varma, R. S. Core–shell Nanoparticles: Synthesis and Applications in Catalysis and Electrocatalysis. *Chem. Soc. Rev.* **2015**, *44*(21), 7540–7590.

Geveke, D. J.; Bigley, A. B. W.; Brunkhorst, C. D. Pasteurization of Shell Eggs using Radio Frequency Heating. *J. Food Eng.* **2017**, *193*, 53–57.

Gouado, I.; Demasse, M. A.; Etame, L. G.; Meyimbo, O.; Ruphine, S.; Ejoh, A.; Fokue, E. Impact of Three Cooking Methods (Steaming, Roasting on Charcoal and Frying) on the β-Carotene and Vitamin C Contents of Plantain and Sweet Potato. *Amer. J. Food Technol.* **2011**, *6*(11), 994–1001.

Guo, Q.; Piyasena, P.; Mittal, G. S.; Si, W.; Gong, J. Efficacy of Radio Frequency Cooking in the Reduction of *Escherichia coli* and Shelf Stability of Ground Beef. *Food Microbiol.* **2006**, *23*(2), 112–118.

Guo, Q.; Sun, D. W.; Cheng, J. H.; Han, Z. Microwave Processing Techniques and their Recent Applications in the Food Industry. *Trends Food Sci. Technol.* **2017**, *67*, 236–247.

Halden, K.; De Alwis, A. A.; Fryer, P. J. Changes in the Electrical Conductivity of Foods During Ohmic Heating. *Int. J. Food Sci. Technol.* **1990**, *25*, 9–35.

Hornig, S.; Heinze T.; Remzi, C. B.; Schubert, U. S. Synthetic Polymeric Nanoparticles by Nanoprecipitation. *J. Mater. Chem.* **2009**, *19*(23), 3838–3840

Horsak, I.; Slama, I. Computer Construction of Ternary Phase Diagrams. *Chem. Papers* **1987**, *41*, 23–33.

Hradecky, J.; Kludska, E.; Belkova, B.; Wagner, M.; Hajslova, J. Ohmic Heating: A Promising Technology to Reduce Furan Formation in Sterilized Vegetable and Vegetable/Meat Baby Foods. *Innov. Food Sci. Emerg. Technol.* **2017**, *43*, 1–6.

Hu, W. T.; Guo, W. L.; Meng, A. Y.; Sun, Y.; Wang, S. F.; Xie, Z. Y. A Metabolomic Investigation into the Effects of Temperature on Streptococcus Agalactiae from Nile tilapia (Oreochromis niloticus) Based on UPLC–MS/MS. *Vet. Microbiol.* **2017**, *210*, 174–182.

Jiang, H.; Wang, S. Highly Efficient Vegetable Drying Technology I: Microwave and Radio Frequency Drying of Vegetables. In *Handbook of Drying of Vegetables and Vegetable Products*, 1st ed.; Zhang, M., Bhandari, B., Fang, Z., Eds.; CRC Press, 2017; pp 45–63.

Kanjanapongkul, K. Rice Cooking using Ohmic Heating: Determination of Electrical Conductivity, Water Diffusion and Cooking Energy. *J. Food Eng.* **2017,** *192,* 1–10.

Kaur, N.; Singh, A. K. Ohmic Heating: Concept and Applications—a Review. *Crit. Rev. Food Sci. Nutr.* **2016,** *56,* 2338–2351.

Kim, S. S.; Kang, D. H. Combination Treatment of Ohmic Heating with Various Essential Oil Components for Inactivation of Food-Borne Pathogens in Buffered Peptone Water and Salsa. *Food Cont.* **2017,** *80,* 29–36.

Knoerzer, K.; Regier, M.; Schubert, H. Measuring Temperature Distributions During Microwave Processing. In *The Microwave Processing of Foods*; Regier, M., Knoerzer, K., Schubert, H., Eds.; Woodhead Publishing: Cambridge, 2017; pp 23–43.

Koray, P. T.; Miran, W. Experimental Comparison of Microwave and Radio Frequency Tempering of Frozen Block of Shrimp. *Innov. Food Sci. Emerg. Technol.* **2017,** *41,* 292–300.

Koubaa, M.; Mhemdi, H.; Fages, J. Recovery of Valuable Components and Inactivating Microorganisms in the Agro-Food Industry with Ultrasound-Assisted Supercritical Fluid Technology. *J. Supercrit. Fluids* **2018,** *134,* 71–79.

Kumar, C.; Saha, S. C.; Sauret, E.; Karim, A.; Gum, Y. Mathematical Modelling of Heat and Mass Transfer During Intermittent Microwave-Convective Drying (IMCD) of Foods. In School of Chemistry; Physics and Mechanical Engineering; Queensland University of Technology Proceedings of the 10th Australasian heat and mass transfer Conference Brisbane, Australia, July 14–15, 2016, 171–176.

Kumar, M.; Jyoti; Hausain, A. Effect of Ohmic Heating of Buffalo Milk on Microbial Quality and Tissue of Paneer. *J. Dairy Foods Home Sci.* **2014,** *33,* 9–13.

Kumar-Singh, S.; Sai-Pavan, M.; Sai-Prasanna, N.; Kant, R. Applications of Super Critical Fluid Extraction in Milk and Dairy Industry: A review. *J. Food. Process. Technol.* **2018,** 9(12), 769.

Kuo, F. J.; Sheng, C. T.; Ting, C. H. Evaluation of Ultrasonic Propagation to Measure Sugar Content and Viscosity of Reconstituted Orange Juice. *J. Food Eng.* **2008,** *86,* 84–90.

Lebovka, N. I.; Shynkaryk, M. V.; Vorobiev, E. Drying of Potato Tissue Pretreated by Ohmic Heating. *Drying Technol.* **2006,** *24,* 601–608.

Leizerson, S.; Shimoni, E. Effect of Ultrahigh-Temperature Continuous Ohmic Heating Treatment on Fresh Orange Juice. *J. Agric. Food Chem.* **2005,** *53,* 3519–3524.

Lévai, G., Martín, Á.; Rodríguez-Rojo, S., Cocero, M. J. Efficient Production of Soy-Bean Lecithin–Pluronic L64® Encapsulated Quercetin Particles in Nanometric Scale using SFEE and PGSS Drying Processes. 5th International Congress on Green Process Engineering (GPE 2016). June 19–24, 2016. Fairmont Tremblant Hotel, Mont Tremblant, Quebec. ECI - Digital archives.

Liu, Z. L.; Wang, X.; Yao, K. L.; Du, G. H.; Lu, Q. H.; Ding Z. H.; Tao, J.; Ning, Q; Luo, X. P.; Tian D. Y.; Xi, D. Synthesis of Magnetite Nanoparticles in W/O Microemulsion. *J. Mater. Sci.* **2004,** *39*(7), 2633–2636.

Yvan, L.; Liu, S.; Fukuoka, M.; Sakai, N. Computer Simulation of Radiofrequency Defrosting of Frozen Foods. *J. Food Eng.* **2015,** *152,* 32–42.

Loghavi, L.; Sastry, S. K.; Yousef, A. E. Effect of Moderate Electric Field Frequency and Growth Stage on the Cell Membrane Permeability of Lactobacillus Acidophilus. *Biotechnol. Prog.* **2009,** *25,* 85–94.

Lu, A. H.; Salabas, E. L.; Schüth, F. Magnetic Nanoparticles: Synthesis, Protection, Functionalization, and Application. *Angew. Chem.* **2007,** *46*(8), 1222–1244.

Lucia, O.; Maussion, P.; Dede, R. J.; Burdio, J. M. Induction Heating Technology and Its Applications: Past Developments, Current Technology, and Future Challenges. *IEEE Trans. Indus. Elect.* **2014**, *61*(5), 2509–2520.

Lyu, X.; Peng, X.; Wang, S.; Yang, B.; Wang, X.; Yang, H.; Xiao, Y.; Baloch, A. B.; Xia, X. Quality and Consumer Acceptance of Radio Frequency and Traditional Heat Pasteurised Kiwi Puree During Storage. *Int. J. Food Sci. Technol.* **2018**, *53*(1), 209–218.

Malik, M. A.; Mohammad, Y. W.; Mohd, A. H. Microemulsion Method: A Novel Route to Synthesize Organic and Inorganic Nanomaterials. 1st Nano Update. *Arab. J. Chem.* **2012**, *5*(4), 397–417.

Manzocco, L.; Anese, M.; Nicoli, M. C. Radiofrequency Inactivation of Oxidative Food Enzymes in Model Systems and Apple Derivatives. *Food Res. Int.* **2008**, *41*(10), 1044–1049.

Mamvura, T. A.; Iyuke, S. E.; Paterson, A. E. Energy Changes During use of High-Power Ultrasound on Food Grade Surfaces. *South Afric. J. Chem. Eng.* **2018**, *25*, 62–73.

Marcotte, M.; Taherian, A. R.; Karimi, Y. Thermophysical Properties of Processed Meat and Poultry Products. *J. Food Eng.* **2008**, *88*(3), 315–322.

Maric, M.; Grassino, A. N.; Zhu, Z.; Barba, F. J.; Brincic, M.; Brincic, S. R. An Overview of the Traditional and Innovative Approaches for Pectin Extraction from Plant Food Wastes and By-Products: Ultasound-; Microwaves-; and Enzyme-Assisted Extraction. *Trends Food Sci. Technol.* **2018**, *76*, 28–37.

Martínez, R. C.; Mohamad, T.; Waisudin, B.; Karim, M.; Hélène, G. G.; Qand, A. N.; Galindo, R. S. A; Álvarez, R. R.; Hatem, F.; Abdelhamid, E. Nanoprecipitation Process: From Encapsulation to Drug Delivery. *Int. J. Pharm.* **2017**, *532*(1), 66–81.

Maric, M.; Grassino, A. N.; Zhu, Z.; Barba, F. J.; Brincic, M.; Brincic, S. R. An Overview of the Traditional and Innovative Approaches for Pectin Extraction from Plant Food Wastes and By-Products: Ultrasound-; Microwaves-; and Enzyme-Assisted Extraction. *Trends Food Sci. Technol.* **2018**, *76*, 28–37.

Markom, M.; Hasan, M.; Daud, W. R. W.; Anuar, N.; Hassan, O.; Singh, H. Chemical Profiling and Quantification of Tannins in *phyllanthusnirurilinn*. Fractionated by Safe Method. *Separation Sci. Technol.* **2010**, *46*, 71–78.

Mélinon, P.; Begin-Colin, S.; Duvail, J. L.; Gauffre, F.; Boime, N. H.; Ledoux, G.; Warot-Fonrose, B. Engineered Inorganic Core/Shell Nanoparticles. *Phys. Rep.* **2014**, *543*(3), 163–197.

Mensink, M. A.; Henderik W. F.; Kees, V. M.; Wouter, L. J. H. How Sugars Protect Proteins in the Solid State and During Drying (Review): Mechanisms of Stabilization in Relation to Stress Conditions. *Eur. J. Pharm. Biopharm.* **2017**, *114*, 288–295.

Michael, M.; Phebus, R. K.; Thippareddi, H.; Subbiah, J.; Birla, S. L.; Schmidt, K. A. Validation of Radio-Frequency Dielectric Heating System for Destruction of *Cronobacter sakazakii* and *Salmonella* Species in Nonfat Dry Milk. *J. Dairy Sci.* **2014**, *97*(12), 7316–7324.

Montañés, F.; Tallon, S. Supercritical Fluid Chromatography as a Technique to Fractionate High-Valued Compounds from Lipids. *Separations* **2018**, *5*, 38.

Moraga, N.; Torres, A.; Guarda, A.; Galotto, M. J. Non-Neutonian Canned Liquid Floor, Unsteady Fluid Mechanics and Heat Transfer Prediction for Pasteurization and Sterilization. *Int. J. Food Eng.* **2011**, *34*, 2000–2025.

Murugesan, R.; Valérie, O. Spray Drying for the Production of Nutraceutical Ingredients-A Review. *Food Bioproc. Technol.* **2012**, *5*(1), 3–14.

Nagaraj, G.; Purohit, A.; Harrison, M.; Singh, R.; Hung, Y. C.; Mohan, A. Radiofrequency Pasteurization of Inoculated Ground Beef Homogenate. *Food Control* **2016**, *59*, 59–67.

Nazim-Cifti, O. Supercritical Fluid Technology: Application to Food Processing. *J. Food Process. Technol.* **2012**, *3*, 5.

Nimratbir, K.; Singh, A. K. Ohmic Heating: Concept and Applications- A Review. *Crit. Rev. Food Sci. Nutr.* **2015**, *56*(14), 2338–2351.

Ojha, K. S.; Mason, T. J.; O′Donnell, C. P.; Kerry, J. P.; Tiwari, B. K. Ultrasound Technology for Food Fermentation Applications. *Ultrasoin. Sonochem.* **2017**, *34*, 410–417.

Ozkoc, S. O.; Sumnu, G.; Sahin, S. Recent Developments in Microwave Heating. Emerging Technologies for Food Processing, 2nd ed.; Academic Press Elsevier, 2014; pp 361–383.

Paniwnyk, L. Applications of Ultrasound in Processing of Liquid Foods: A Review. *Ultrason. Sonochem.* **2017**, *38*, 794–806.

Parhi, R.; Suresh, P. Supercritical Fluid Technology: A Review. *J. Adv. Pharm. Sci. Technol.* **2012**, *1*(1), 1–13.

Pereira, R.; Martins, J.; Mateus, C.; Teixeira, J. A.; Vicente, A. A. Death Kinetics of *Escherichia coli* in Goat Milk and *Bacillus licheniformis* in Cloudberry Jam Treated by Ohmic Heating. *Chem. Pap.* **2007**, *61*, 121–126.

Pinela, J.; Prieto, M. A.; Pereira, E.; Jabeur, I.; Barreiro, M. F.; Barros, L.; Ferreira, I. C. F. R. Optimization of Heath- and Ultrasound-Assisted Extraction of Anthocyanins from Hibiscus sabdariffa calyces for Natural Food Colorants. *Food Chem.* **2019**, *275*, 309–321.

Pingret, D.; Fabiano-Tixier, A. S.; Chemat, F. Degradation During Application of Ultrasound in Food Processing: A review. *Food Control* **2013**, *31*, 593–606.

Rapoport, E.; Pleshivtseva, Y. Introduction to Theory and Industrial Application of Induction Heating Processes. In *Optimal Control of Induction Heating Processes,* 1st ed.; CRC Press, 2006; pp 1–34.

Rozzi, N. L.; Singh, R. K. Supercritical Fluids and the Food Industry. *Compr. Rev. Food Sci. Food Saf.* **2002**, *1*, 33–44.

Ryang, J. H.; Kim, N. H.; Lee, B. S.; Kim, C. T.; Rhee, M. S. Destruction of *Bacillus cereus* Spores in a Thick Soy Bean Paste (doenjang) by Continuous Ohmic Heating with Five Sequential Electrodes. *Lett. Appl. Microbiol.* **2016**, *63*, 66–73.

Ty′skiewicz, K.; Konkol, M.; Rój, E. The Application of Supercritical Fluid Extraction in Phenolic Compounds Isolation from Natural Plant Materials. *Molecules* **2018**, *23*, 2625.

Sagong, H. G.; Park, S. H.; Choi, Y. J.; Ryu, S.; Kang, D. H. Inactivation of *Escherichia coli* O157:H7, *Salmonella typhimurium*, and *Listeria monocytogenes* in Orange and Tomato Juice Using Ohmic Heating. *J. Food Protect.* **2011**, *74*, 899–904.

Sakr, M.; Liu, S. A Comprehensive Review on Applications of Ohmic Heating (OH). *Renew. Sustain. Energy Rev.* **2014**, *39*, 262–269.

Santana, F.; Duarte, P. E.; Cristianini, M. Determination of the Convective Heat Transfer Coefficient (*h*) in the Sterilization of Retortable Pouches. *Int. J. Food Eng.* **2011**, *7*(1), 3–14.

Salgado, M.; Rodríguez-Rojo, S.; Alves-Santos, F. M.; Cocero, M. J. Encapsulation of Resveratrol on Lecithin and β-Glucans to Enhance its Action Against *Botrytis cinerea. J. Food Eng.* **2015**, *165*, 13–21.

Sarang, S.; Sastry, S. K.; Knipe, L. Electrical Conductivity of Fruits and Meats During Ohmic Heating. *J. Food Eng.* **2008**, *87*(3), 351–356.

Sasson, A.; Monselise, S. P. Electrical Conductivity of Shamouti Orange Peel During Fruit Growth and Postharvest Senescence. *J. Amer. Soc. Horticult. Sci.* **1977**, *102*, 142–144.

Singh, S. B.; Tandon P. K. Catalysis: A Brief Review on Nano-Catalyst. *J. Ener. Chem. Eng.* **2014**, *2*(3), 106–115.

Solans, C.; García-Celma, M. J. Microemulsions and Nano-Emulsions for Cosmetic Applications. Cosmetic Science and Technology. *Theor. Princ. Appl.* **2017**, *11*, 507–518.

Sosa-Morales, M. E.; Tiwari, G.; Wang, S.; Tang, J.; Garcia, H. S.; Lopez-Malo, A. Dielectric Heating as a Potential Post-Harvest Treatment of Disinfesting Mangoes, Part II: Development of Rf-Based Protocols and Quality Evaluation of Treated Fruits. *Biosyst. Eng.* **2009**, *103*(3), 287–296.

Tang, J.; Hong, Y. K.; Inanoglu, S.; Liu, F. Microwave Pasteurization for Ready-to-Eat Meals. *Cur. Opin. Food Sci.* **2015**, *23*, 133–141.

Tian, X. J.; Yu, Q. Q.; Shao, L. L.; Li, X. M.; Dai, R. T. Sublethal Injury and Recovery of Escherichia coli O157:H7 After Ohmic Heating. *Food Cont.* **2018**, *94*, 85–92.

Uemura, K.; Kanafusa, S.; Takahashi, C.; Kobayashi, I. Development of a Radio Frequency Heating System for Sterilization of Vacuum-Packed Fish in Water. *Biosci. Biotechnol. Biochem.* **2017**, *81*(4), 762–767.

Valle, J. M., Fuente, J. C.; Cardarelli, D. A. Contributions to Supercritical Extraction of Vegetable Substrates in Latin America. *J. Food Eng.* **2005**, *67*, 35–57.

Xiaojing, T.; Qianqian, Y.; Donghao, Y.; Lele, S.; Zhihong, L., Fei, J.; Xingmin, Li.; Teng, H.; Ruitong, D. New Insights into the Response of Metabolome of *Escherichia coli* O157:H7 to Ohmic Heating. *Front. Microbiol.* **2018**, *9*, 2936.

Xiaojing, T.; Qianqian, Y.; Wei, W.; Ruitong, D. Inactivation of Microorganisms in Foods by Ohmic Heating: A Review. *J. Food Protec.* **2018**, *81*(7), 1093–1107.

Xu, J.; Zhang, M.; Bhandari, B.; Kachele, R. ZnO Nanoparticles Combined Radio Frequency Heating: A Novel Method to Control Microorganism and Improve Product Quality of Prepared Carrots. *Innov. Food Sci. Emerg. Technol.* **2017**, *44*, 46–53.

Yang, C.; Zhao, Y.; Tang, Y.; Yang, R.; Yan, W.; Zhao, W. Radio Frequency Heating as a Disinfestation Method Against *Corcyra cephalonica* and its Effect on Properties of Milled Rice. *J. Stored Prod. Res.* **2018**, *77*, 112–121.

Yao, Y. Enhancement of Mass Transfer by Ultrasound: Application to Adsorbent Regeneration and Food Drying/Dehydration. *Ultrason. Sonochem.* **2016**, *31*, 512–531.

Zareifard, M. R.; Ramaswamy, H. S.; Trigui M.; Marcotte, M. Ohmic Heating Behaviour and Electrical Conductivity of Two-Phase Food Systems. *Innov. Food Sci. Emerg. Technol.* **2003**, *4*, 45–55.

Zell, M.; Lyng, J. G.; Cronin, D. A.; Morgan, D. J. Ohmic Cooking of Whole Beef Muscle Evaluation of the Impact of a Novel Rapid Ohmic Cooking Method on Product Quality. *Meat Sci.* **2010**, *86*, 258–263.

Zell, M.; Lyng, J. G.; Morgan, D. J.; Cronin, D. A. Minimizing Heat Losses During Batch Ohmic Heating of Solid Food. *Food Bioprod. Proc.* **2011**, *89*, 128–134.

Zhan, B.; Xi, M.; Bruschweiler-Li, L.; Brüschweiler, R. Nanoparticle-Assisted Metabolomics. *Metabol.* **2018**, *8*, 1–14.

Zheng, A.; Zhang, L.; Wang, S. Verification of Radio Frequency Pasteurization Treatment for Controlling *Aspergillus parasiticus* on Corn Grains. *Int. J. Food Microbiol.* **2017**, *249*, 27–34.

CHAPTER 13

Cold Plasma: Application in Food Packages

CATALINA J. HERNÁNDEZ-TORRES[1], CRISTÓBAL N. AGUILAR[1*],
MÓNICA L. CHÁVEZ-GONZÁLEZ[1], JOSÉ L. MARTÍNEZ-HERNÁNDEZ[2],
MIRIAM D. DÁVILA-MEDINA[3], and YADIRA K. REYES-ACOSTA[1*]

[1]*Bioprocesses & Bioproducts Group, Food Research Department,
School of Chemistry, Autonomous University of Coahuila,
Saltillo 25280, Coahuila, México*

[2]*Nanobioscience group, Chemistry School, Autonomous University of
Coahuila, Blvd. V. Carranza e Ing. José Cárdenas Valdés, Saltillo 25280,
Coahuila, Mexico*

[3]*Group of bioprocess and microbial biochemistry, School of Chemistry,
Autonomous University of Coahuila, Saltillo 25280, Coahuila, México*

**Corresponding author. E-mail: ykreyes@uadec.edu.mx;
cristobal.aguilar@uadec.edu.mx*

ABSTRACT

Cold plasma (CP) has been defined as an emerging technology that has
gained the attention of the food industry due to its great results inactivating
microorganisms, like fungi, bacteria, viruses, and spores. CP has been
applied to different food like beef, chicken, fruits, vegetables among others,
and has been reported its great results to inactivate microorganisms. Even
though microorganism deactivation before the treatment has been positive,
there has been an interest in applying this technology in food packages to
extend the shelf life of the food once it is out in the market and to preserve
it fresh. Different researchers have reported that after CP treatment in food
packages it has been shown that it is possible to preserve food during more
time without changing the quality of the food treated. The use of different

gases during the CP treatment to carry out different objectives have been reported, some of them are oxygen (O), hydrogen peroxide (H_2O_2), ozone (O_3), nitrogen (N). However, it is important to know that to produce CP, the way the energy is input to the gas is important, these can be carried out by using, barrier discharge, corona, jet configurations, operations parameters such as frequency, voltage, power density, among others. In recent years due to the contamination of the environment, the need to replacing or reuse different materials like glass, metals, and plastic has grown. Different types of materials such as polypropylene, low-density polyethylene, clamshell container, high barrier cryovac, among others, have been used as food packages modified with CP. The use of different materials used in the application of plasma has gained interest but steel has a lot of unknown questions to be solved.

13.1   INTRODUCTION

It is important for the consumer to select the products that met their needs. They usually look for healthy food rich in nutrients, with a great physical aspect, a low price, and a good quality. Considering the increase of the commercial offer in different formats, the way of choosing and buying products or services has changed. Different authors have established that since the consumers have more requirements when buying food it becomes necessary for companies to find ways to improve their productivity in terms of using sustainable materials, safety products, and implementing flexible and standardized technology (Rodríguez-Rojas et al., 2019). Although organizations face new challenges every day as offering better services for customer satisfaction, there is still a lot of work due to the consumer demand for more natural and safe food. Minimize food deterioration before it reaches the consumer is one of the world challenges that require different efforts to focus on developing appropriate packaging systems to extend the food shelf life (Muratore et al., 2019). Polymers are the most common material for food packaging, the main purpose of a package is to be able to protect food from external contaminants and ensure durable storage. However, the use of different substances during the process of a food package, like additives, plasticizers, or other unintentional chemical substances could migrate into food, resulting in dietary exposure of the consumer to chemicals possibly detrimental to human health (Wang et al., 2019). Food packages can be classified into three groups depending on their function, traditional packages purpose is protection, containment, and communication, by the other hand the purpose of an active package (AP) is to reduce or inhibit microbial growth interacting directly with the food products extending their shelf life and intelligent packaging communicates quality and safety information to the consumer before they buy the product (Batista et al., 2019).

Even though, there are different technologies to inactivate microorganisms from food products to extend their shelf life, one technology that has gained much attention to different industries, like the food industry, is CP, which has shown great results in inactivating different microorganism in food. However, this technology has been used in recent years not only in food products but also in food packages, with great results maintaining the characteristics in the packages preserving for more time food products. This chapter entails different researches were CP has been used in food package to inactivate microorganisms to extend the shelf life of the product.

13.2   FOOD PACKAGE

The high demand of the consumer for fresh food with high quality has grown in the last years due to the requirement of better health, to satisfy this need analysis of food safety and packaging has been studied by different authors. Since human population has been growing in the past decades, the need of food has grown more and more, knowing that the food production in the world is not enough to feed the growth generated of global human population. Different alternatives have been searched, one of them is a food packaging that conserves and facilitates the commercialization of the product since is required. According to researchers, food packaging is used for several purposes, for the consumer to have the necessary information about the product, to contain the product, appropriate presentation of the product, define the quality that the consumer buys, to make the transportation easier, and to protect the products from pollution, contamination, or damage (Hamad et al., 2018; Rodríguez-Rojas et al., 2019; Santeramo et al., 2018).

Food contamination can occur during the different stages it goes through, like harvesting, food processing, and distribution (Malhotra et al., 2015; Mousavi et al., 2018). Packaging is an effective way to protect different types of food products from changes like chemical, physical, and biological, during storage or external contaminants. However, the lifestyle that has been generated in recent years has brought us to look for prepared foods like microwave meals to make our life's much easier, thus leading to the importance to look for different food packages and technologies not only to get a ready to eat meal but also to protect the food from microorganisms, external changes, and to preserve food with its original nutrients and benefits during its shelf life (Majid et al., 2018).

Different material has been used like food package, metallic cans, aseptic packaging, flexible packaging, aluminum foil, polyester, polypropylene, glass, paper, hydrogel among others. Moreover, in the 20th century more advancement in packaging technology appeared; one example is the AP. The primary objective of an AP is to maintain and increase the shelf life of food products and they can be considered one of the most innovative ways to inhibit different microbial growth on foods while their shelf life maintaining quality, freshness, and safety (Batista et al., 2019; Erdohan et al., 2013; Mastromatteo et al., 2010; Takma and Korel, 2019; Khaneghah et al., 2018). One alternative that has great benefits is the use of biodegradable AP with natural extracts from plants that have been reported to have antioxidant or antimicrobial. This can be considered by the food industry to improve their products during the different stages it has (Erdohan et al., 2013; Khaneghah et al., 2018; Mastromatteo et al., 2010; Muriel-galet et al., 2013; Muriel-galet et al., 2012; Shemesh et al., 2014; Takma and Korel, 2019).

Another compound used in the AP is nanoparticles which are incorporated with polymer matrices to improve the surface of the package and make it antimicrobial. An example of the nanostructured antimicrobial systems is silver nanoparticles which different authors have reported to have antimicrobial activity against gram-negative and gram-positive bacteria; they form pits and disruption of cell membranes. ZnO is one of the most used metal oxides, due to its remarkable antimicrobial properties, it's recognized as safe (GRAS) by the US Food and Drug Administration (FDA, 2015). ZnO NPs have been reported as effective antimicrobials when they are incorporated into different materials such as poly (3-hydroxybutyrate-co-3-hydroxyvalerate), bacterial cellulose, polycarbonate, fish protein isolate/gelatin, among others (Azeredo et al., 2018; Dhapte et al., 2015).

In the food industry, not only are packaging complements being sought, but also technologies that ensure better food preservation, quality, and benefits. One technology that has been used in food industry and has great results is CP wish has used to modify food packages to preserve for more time the quality, nutrients, and other aspects important for the consumer. Since it has been used to inactivate microorganisms most commonly found in the food industry, it can benefit even better and for more time the product if it's used not only in the food but in the food package.

13.3 PLASMA

Plasma is the fourth state of matter based on the ionization of gas molecules to create reactive chemical species with antimicrobial properties (Min et al., 2018). CP has been used for inactivation of microorganisms in apples (Tappi et al., 2017), lettuce (Min et al., 2017), carrots (Schnabet et al., 2015), tomatoes (Min et al., 2018), blueberries (Lacombe et al., 2015), eggshells (Dasan et al., 2018), black pepper (Mošovská et al., 2018), almonds (Hertwig et al., 2017), meat (Misra and Jo, 2017), fish products (Albertos et al., 2017). It has been used in the food industry for the inactivation of microorganisms most commonly found in food such as *Escherichia coli* (Segura-ponce et al., 2018), *Salmonella* (Timmons et al., 2018), and *Listeria monocytogenes* (Bauer et al., 2017), and inactivation of aflatoxigenic spores of *Aspergillus flavus* and *Aspergillus parasiticus* on hazelnuts (Dasan et al., 2017).

13.4 COLD PLASMA APPLIED TO FOOD PACKAGE

When talking about CP there is a misunderstanding since it has been established that the technology is at low temperatures, but it does not refer to temperatures below zero, nor is there an established temperature range for this technology, it is known that plasma treatments that operate at temperatures close to 60°C are considered CPs. Since food packaging has become important in the food conservation, alternatives have been searched to improve packaging and preserve food better. According to different authors, CP has been used to modify different materials used in food packages, obtaining positive results (Misra et al., 2019a).

Pankaj et al. (2014), reported that after plasma treatment the roughness of polylactic acid (PLA) film increased mainly at the site in contact with high voltage electrode at both the voltage levels of 70 and 80 kV they used. The treatment with plasma did not induce any change in the glass transition temperature, but significant increase in the initial degradation temperature and maximum degradation temperature was observed. It is important to highlight that plasma treatment did not affect the oxygen and water vapor permeability of PLA.

Misra et al. (2014) decontaminated strawberries inside a sealed package with two different gas mixtures, 65% O_2 + 16% N_2 + 19% CO_2, and 90% N_2 + 10% O_2. The microflora presented in the strawberries was reduced by an average of ~3.0 log cycles from the initial levels of 5 $\log_{10}$ CFU/g in 300s of in-package atmospheric CP discharge. Plasma treatments with the two gas mixtures showed similar effects on microbial reduction levels. After plasma treatment strawberries, which were stored, showed favorable quality results with similar respiration rates and an 11% higher firmness than the control stored for 24 h.

However, atmospheric dielectric barrier discharges cold plasma treatment (ADCPT) has gained attention in the food industry due to the generation of CP inside a sealed container that prevents post-process contamination of the treated food. The effectiveness of the ADCPT is influenced by the composition and quantity of the reactive species as well as the possibility of contact between the reactive species and food both of which can be influenced in turn by different packaging parameters like packaging material, package headspace composition, and others. Kimet al. (2019) used atmospheric dielectric barrier discharge cold plasma treatment (ADCPT) to inactivate *Salmonella* in mixed vegetables (carrots, roman lettuce, red cabbage, red tomatoes). Different types of packaging material were used in this study low-density polyethylene (LDPE), composite of nylon (3%)-LDPE were used to create flexible pouch packages. LDPE, polypropylene, and polyethylene terephthalate films were used to construct rigid containers. They reported that after the treatment was observed a reduction of Salmonellain the grape tomatoes treated in the flexible pouch made of the nylon-LDPE composite than the tomatoes treated in the LDPE pouch, with the levels being 2.3 ± 0.1 and 1.8 ± 0.3 log CFU/tomato, respectively. They concluded that reactive species in CP, including reactive oxygen species (ROS), reactive nitrogen species (RNS), other atoms, radicals, and excited molecules, which are diffused through cell membranes, can inactivate microorganisms by reacting with different membranes components like lipids, proteins, enzymes, and nucleic acids, and that the inactivation rate of indigenous aerobes of grape tomatoes had better results in low-density polyethylene packaging.

Prasad et al. (2017) treated tomatoes inoculated with E. coli at a concentration of 10^8 log CFU mL$^-$1 and treated with cold plasma (ACP). The evaluations were carried out at room temperature (25°C) and temperatures of refrigeration (4°C) during 48 h. The inoculated tomatoes were exposed ACP at 15 and 60 kV for 5, 10, 15, and 30 min followed by their storage at 4°C and 25°C. The best reduction treatment was obtained during the ACP treatment of 60 kV for 15 min with a reduction of 6 log CFU mL^{-1}. They concluded that ACP is a potential technique for fruits and vegetables that are eaten raw like tomatoes.

On the other hand, Misra et al. (2014) used a dielectric barrier discharge (DBD) which generates a plasma between two electrodes at the high potential difference, separated by one or more dielectric barriers. That is one of the benefits of the use of such type of plasma. The authors reported the use of noble gases such as argon, nitrogen, oxygen, and/or air for attaining the plasma state. However, it is common to use various gas mixtures in food packages for extension of shelf life and quality, it is called modified atmosphere packaging (MAP) which refers to any atmosphere different from normal.

When using any kind of plasma, the reactions are very complex due to a mixture of multi-atomic molecules in the feed gas to plasma, another complexity is due to the water molecules present in the gas or the ones introduced from food materials placed in the discharge. Considering these, it is obvious that humid air plasma is by far the most complex in terms of the chemistry. The air plasma is consider a potent source of a myriad of ROS that in humid air plasma with antimicrobial activity include hydrogen peroxide (H_2O_2), ozone (O_3), superoxide anion ($O_2(-)$), hydroperoxyl (HO_2), alkoxyl (RO-), peroxyl, singlet oxygen (O_2), hydroxyl radical ($\bullet OH$), and carbonate anion radical ($CO_3 \bullet -$). RNS, for example, nitric oxide (NO), nitrogen dioxide radical (NO_2), peroxynitrite ($ONOO-$), peroxynitrous acid (OONOH), and alkylperoxynitrite (ROONO). The reactive species and their concentration in the plasma would vary depending on many factors, including the gas in which plasma is induced, the configuration of the plasma source, power input to the gas, duration of treatment, and the humidity levels (Misra and Jo, 2017; Misra et al., 2019[b]; Prasad et al., 2017).

13.5   CONCLUSION

Food package modified with CP is a new alternative for preserving food from different types of microorganisms during transportation or until it gets to the consumers' table. CP has been reported to be capable of inactivating microorganisms in foods including bacteria, spores, fungi, yeast, and viruses, all of these in different kinds of food including fruits and vegetables, fish, meat, among others. There're different types of mechanisms and gases to generate CP, from oxygen, nitrogen wish interacts during the process of CP depending on the type of plasma that has been used, atmospheric CP, DBD. For in-package plasma processes, it is essential to ensure that besides general food grade requirements for packaging materials, the ability to withstand the plasma discharges without causing chemical migration or pinholes are met.

KEYWORDS

- **cold Plasma**
- **technological application**
- **antimicrobial properties**
- **food control**
- **food packages**

REFERENCES

Albertos, I.; Martín-Diana, A. B.; Cullen, P. J.; Tiwari, B. K.; Ojha, S. K.; Bourke, P.; Rico, D. Effects of dielectric Barrier Discharge (DBD) Generated Plasma on Microbial Reduction and Quality Parameters of Fresh Mackerel (Scomber scombrus) fillets. *Innov. Food Sci. Emerg. Technol.* **2017**. https://doi.org/10.1016/j.ifset.2017.07.006

Azeredo, H. M. C.; de Otoni, C. G.; Assis, O. B. G.; Corrêa, D. S.; de Moura, M. R.; Mattoso, L. H. C. Nanoparticles, and Antimicrobial Food Packaging. *Ref. Module Food Sci.* **2018**, 1–13.

Batista, R. A.; Espitia, P. J. P.; Quintans, J.; Freitas, M. M.; Cerqueira, M. Â.; Teixeira, J. A.; Cardoso, J. C. Hydrogel as an Alternative Structure for Food Packaging Systems. *Carbohydr. Polym.* **2019**, *205*, 106–116.

Bauer, A.; Ni, Y.; Bauer, S.; Paulsen, P.; Modic, M.; Walsh, J. L.; Smulders, F. J. M. The Effects of Atmospheric Pressure Cold Plasma Treatment on Microbiological, Physical-Chemical and Sensory Characteristics of Vacuum Packaged Beef Loin. *Meat Sci.* **2017**, *128*, 77–87.

Dasan, B. G.; Boyaci, I. H.; Mutlu, M. Nonthermal Plasma Treatment of Aspergillus spp. Spores on Hazelnuts in an Atmospheric Pressure Fluidized Bed Plasma System: Impact of Process Parameters and Surveillance of the Residual Viability of Spores. *J. Food Eng.* **2017**, *196*, 139–149.

Dasan, B. G.; Yildirim, T.; Boyaci, I. H. Surface Decontamination of Eggshells by Using Non-Thermal Atmospheric Plasma. *Int. J. of Food Microbiol.* **2018**, *266*, 267–273.

Dhapte, V.; Gaikwad, N.; More, P. V.; Banerjee, S.; Dhapte, V. V.; Kadam, S.; Khanna, P. K. Transparent ZnO/Polycarbonate Nanocomposite for Food Packaging Application. *Nanocomposites* **2015**, *1*(2), 106–112.

Erdohan, Z. O.; Çam, B.; Nazan, K. Characterization of Antimicrobial Polylactic Acid Based Films. *J. Food Eng.* **2013**, *119*, 308–315.

Hamad, A. F.; Han, J. H.; Kim, B. C.; Rather, I. A. The Intertwine of Nanotechnology with the Food Industry. *Saudi J. Biol. Sci.* **2018**, *25*(1), 27–30.

Hertwig, C.; Leslie, A.; Meneses, N.; Reineke, K.; Rauh, C.; Schlüter, O. Inactivation of Salmonella Enteritidis PT30 on the Surface of Unpeeled Almonds by Cold Plasma. *Innov. Food Sci. Emerg. Technol.* 2017, *44*, 242–248.

Khaneghah, A. M.; Hashemi, S. M. B.; Es, I.; Fracassetti, D.; Limbo, S. Efficacy of Antimicrobial Agents for Food Contact Applications : Biological Activity, Incorporation into Packaging, and Assessment Methods : A Review. *J. Food Prot.* **2018**, *81*(7), 1142–1156.

Kim, S. Y.; Bang, I. H.; Min, S. C. Effects of Packaging Parameters on the Inactivation of Salmonella Contaminating Mixed Vegetables in Plastic Packages Using Atmospheric Dielectric Barrier Discharge Cold Plasma Treatment. *J. Food Eng.* **2019,** *242,* 55–67.

Lacombe, A.; Niemira, B. A.; Gurtler, J. B.; Fan, X.; Sites, J.; Boyd, G.; Chen, H. Atmospheric Cold Plasma Inactivation of Aerobic Microorganisms on Blueberries and Effects on Quality Attributes. *Food Microbiol.* **2015,** *46,* 479–484.

Majid, I.; Ahmad Nayik, G.; Mohammad Dar, S.; Nanda, V. Novel Food Packaging Technologies: Innovations and Future Prospective. *J. Saudi Soc. Agric. Sci.* **2018,** *17*(4), 454–462.

Malhotra, B.; Keshwani, A.; Kharkwal, H. Antimicrobial Food Packaging : Potential and Pitfalls. *Front Microbiol.* **2015,** *6*(611), 1–9. https://doi.org/10.3389/fmicb.2015.00611

Mastromatteo, M.; Mastromatteo, M.; Conte, A. Advances in Controlled Release Devices for Food Packaging Applications. *Trends Food Sci. Technol.* **2010,** *21*(12), 591–598.

Min, S. C.; Hyeon, S.; Niemira, B. A.; Boyd, G.; Sites, J. E.; Fan, X.; Jin, T. Z. In-Package Atmospheric Cold Plasma Treatment of Bulk Grape Tomatoes for Microbiological Safety and Preservation. *Food Res. Int.* **2018,** *108,* 378–386.

Min, S. C.; Roh, S. H., Niemira, B. A.; Boyd, G.; Sites, J. E.; Fan, X., Jin, T. Z. In-Package Atmospheric Cold Plasma Treatment of Bulk Grape Tomatoes for Microbiological Safety and Preservation. *Food Res. Int.* **2018,** *108,* 378–386.

Min, S.; Roh, S.; Niemira, B.; Boyd, G.; Sites, J.; Uknalis, J.; Fan, X. In-Package Inhibition of E. coli O157: H7 on Bulk Romaine Lettuce Using Cold Plasma. *Food Microbiol.* **2017,** *65,* 1–6.

Misra, N. N.; Jo, C. Applications of Cold Plasma Technology for Microbiological Safety in Meat Industry. *Trends Food Sci. Technol.* **2017,** *64,* 74–86.

Misra, N. N.; Moiseev, T.; Patil, S.; Pankaj, S. K.; Bourke, P.; Mosnier, J. P.; Cullen, P. J. Cold Plasma in Modified Atmospheres for Post-harvest Treatment of Strawberries. *Food Bioprocess Technol.* **2014,** *7*(10), 3045–3054.

Misra, N. N.; Patil, S.; Moiseev, T.; Bourke, P.; Mosnier, J. P.; Keener, K. M.; Cullen, P. J. In-Package Atmospheric Pressure Cold Plasma Treatment of Strawberries. *J. Food Eng.* **2014,** *125,* 131–138.

Misra, N. N.; Yepez, X.; Xu, L.; Keener, K. In-Package Cold Plasma Technologies. *J. Food Eng.* **2019a,** *244,* 21–31.

Misra, N. N.; Yepez, X.; Xu, L.; Keener, K. In-Package Cold Plasma Technologies. *J. Food Eng.* **2019b.** https://doi.org/10.1016/j.jfoodeng.2018.09.019

Mošovská, S.; Medvecká, V.; Halászová, N.; Ďurina, P.; Valík, Ľ.; Mikulajová, A.; Zahoranová, A. Cold Atmospheric Pressure Ambient Air Plasma Inhibition of Pathogenic Bacteria on the Surface of Black Pepper. *Food Res. Int.* **2018,** *106,* 862–869.

Mousavi Khaneghah, A.; Hashemi, S. M. B.; Limbo, S. Antimicrobial Agents, and Packaging Systems in Antimicrobial Active Food Packaging: An Overview of Approaches and Interactions. *Food Bioprod. Process.* **2018,** *111,* 1–19.

Muratore, F.; Barbosa, S. E.; Martini, R. E. Development of Bioactive Paper Packaging for Grain-Based Food Products. *Food Packag. Shelf Life* **2019,** *20,* 100–317.

Muriel-galet, V.; Cerisuelo, J. P.; López-Carballo, G.; Aucejo, S.; Gavara, R. Evaluation of EVOH-Coated PP Films with Oregano Essential Oil and Citral to Improve the Shelf-Life of Packaged Salad. *Food Control* **2013,** *30*(1), 137–143.

Muriel-galet, V.; Cerisuelo, J. P.; López-Carballo, G.; Lara, M.; Gavara, R.; Hernández-Muñoz, P. Development of Antimicrobial Films for Microbiological Control of Packaged Salad. *Int. J. Food Microbiol.* **2012,** *157*(2), 195–201.

Pankaj, S. K.; Bueno-Ferrer, C.; Misra, N. N.; O'Neill, L.; Jiménez, A.; Bourke, P.; Cullen, P. J. Characterization of Polylactic Acid Films for Food Packaging as Affected by Dielectric Barrier Discharge Atmospheric Plasma. *Innov. Food Sci. Emerg. Technol.* **2014,** *21*, 107–113.

Prasad, P.; Mehta, D.; Bansal, V.; Sangwan, R. S. Effect of Atmospheric Cold Plasma (ACP) With its Extended Storage on the Inactivation of Escherichia coli Inoculated on Tomato. *Food Res. Int.* **2017.** https://doi.org/10.1016/j.foodres.2017.09.030

Rodríguez-Rojas, A.; Arango-Ospina, A.; Rodríguez-Vélez, P.; Arana-Florez, R. ¿What is New About food Packaging Material? A Bibliometric Review From 1996–2016. *Trends Food Sci. Technol.* **2019,** *85*, 252–261.

Santeramo, F. G.; Carlucci, D.; De Devitiis, B.; Seccia, A.; Stasi, A.; Viscecchia, R.; Nardone, G. Emerging Trends in European Food, Diets, and Food Industry. *Food Res. Int.* **2018,** *104*, 39–47.

Schnabel, U.; Niquet, R.; Schlüter, O.; Gniffke, H.; Ehlbeck, J. Decontamination and Sensory Properties of Microbiologically Contaminated Fresh Fruits and Vegetables by Microwave Plasma Processed Air (PPA). *J. Food Process. Preserv.* **2015,** *39*(6), 653–662.

Segura-ponce, L. A.; Reyes, J. E.; Troncoso-Contreras, G.; Valenzuela-Tapia, G. Effect of Low-Pressure Cold Plasma (LPCP) on the Wettability and the Inactivation of Escherichia coli and Listeria innocua on Fresh-Cut Apple (Granny Smith) Skin. *Food Bioprocess Technol.* **2018.**

Shemesh, R.; Goldman, D.; Krepker, M.; Danin-poleg, Y.; Kashi, Y.; Vaxman, A.; Segal, E. LDPE/Clay/Carvacrol Nanocomposites with Prolonged Antimicrobial Activity. *J. Appl. Polym. Sci.* **2014,** *41261*, 1–8.

Takma, D. K.; Korel, F. Active Packaging Films as a Carrier of Black Cumin Essential Oil : Development and Effect on Quality and Shelf-Life of Chicken Breast Meat. *Food Packag. Shelf Life* **2019,** *19*, 210–217.

Tappi, S.; Ragni, L.; Tylewicz, U.; Romani, S.; Ramazzina, I.; Rocculi, P. Browning Response of Fresh-Cut Apples of Different Cultivars to Cold Gas Plasma Treatment. *Innov. Food Sci. Emerg. Technol.* **2017.** https://doi.org/10.1016/j.ifset.2017.08.005

Timmons, C.; Pai, K.; Jacob, J.; Zhang, G.; Ma, L. M. Inactivation of Salmonella enterica, Shiga Toxin-Producing Escherichia coli and Listeria Monocytogenes by a Novel Surface Discharge Cold Plasma Design. *Food Control* **2018,** *84*, 455–462.

Wang, C.; Gao, W., Liang, Y.; Jiang, Y; Wang, Y.; Zhang, Q.; Jiang, G. Migration of Chlorinated Paraffins from Plastic Food Packaging into Food Simulants: Concentrations and Differences in Congener Profiles. *Chemosphere* **2019,** *225*(December 2017), 557–564.

CHAPTER 14

Current Processes of Recovery, Separation, and Purification of Biocompounds with Potential Application in the Food Industry

LILIANA LONDOÑO-HERNANDEZ[1,2*], CRISTINA RAMÍREZ-TORO[2], DIANA A. BRICEÑO-VELEZ[2], and AYERIM HERNÁNDEZ-ALMANZA[3*]

[1]*Biotics Group, School of Basic Sciences, Technology and Engineering, Universidad Nacional Abierta y a Distancia UNAD, Colombia*

[2]*Research Group in Microbiology and Marine Biotechnology - MIBIA. Biology Department, Universidad del Valle, Cali, 25360, Valle del Cauca, Colombia*

[3]*School of Biological Science, Universidad Autónoma de Coahuila, Torreón 27000, Coahuila, México*

Corresponding author. E-mail: ayerim_hernandez@uadec.edu.mx.; lilianalondono@uadec.edu.mx

ABSTRACT

With the increase in agro-industrial processes has increased the generation of low-value by-products or waste, which is generally disposed of inappropriately in landfills or other sites, causing serious environmental damage. These agro-industrial by-products are composed mostly of carbohydrates, proteins, minerals, among others, which could be used to obtain value-added molecules. Therefore, the valorization of these materials becomes a priority, in response to current trends of circular economy that indicate that it is necessary to reuse these by-products, in addition to the economic losses generated by wasting such materials. One of the alternative uses would be to obtain compounds with biological activities of interest such as antioxidants, antihypertensives, anticancer, among others, to be introduced in food matrices, which would allow obtaining foods with functional characteristics, extend the useful life of products, improve sensory characteristics, among others. In recent years, research into the processes of obtaining, separating, and purifying high value-added compounds from agro-industrial by-products has gained a special place in the scientific community. However, most extraction processes are based on the use of solvents, which is contrary to current environmental regulations. Therefore, it is necessary to search for and implement alternative technologies that will efficiently obtain these products. The objective of this one is to know the advances in the technologies used for the recovery of compounds and their application in the revaluation of agro-industrial by-products, discussing some technological and economic aspects, and the limitations for the use of these technologies to scales superior to those of laboratory.

14.1   INTRODUCTION

Currently, most industries generate waste or by-products considered to have low added value. One of the activities that generate the most waste is agro-industry. According to statistical reports, agro-industrial activities generate about 90 million tons of by-products annually, which are generally disposed of improperly, being one of the most important sources of pollution, causing serious environmental problems. The organic residues derived from these activities can be basically divided into two groups, those coming from agriculture and forestry, and those coming from industrial activities. These residues include: livestock manure, crop residues, pruning residues, husks, seeds, stems, leaves, whey, among others (Roselló-Soto et al., 2015; Yusuf, 2017).

In recent years, as a strategy to reduce the effects of pollution on the environment and conserve natural resources, the concept of a circular economy has been established, whose purpose is to maintain the economic value of products and resources longer and reduce waste generation (Melgarejo-Moreno, 2019). In this sense, the use of agro-industrial waste in different sectors has been proposed, mitigating the negative impact that these may have. The latter have characterized various agro-industrial wastes, demonstrating that they are a potential source of vitamins, minerals, proteins, carbohydrates, fibers, bioactive compounds, such as polyphenols, among others, which have various biological properties such as antioxidants, antimicrobials, antihypertensives, anticancerogenic, and so on, which encourages interest in the use of such wastes to obtain high value-added products (Sadh et al., 2018; Roselló-Soto et al., 2019).

Due to its composition, the use of agro-industrial residues in different industries has been reported for the production of molecules or compounds of high-added value such as enzymes, pigments, fertilizers, plant growth promoting molecules, compounds with biological activity of interest, among others (Yusuf, 2017). The use of these products, and especially of compounds with biological activity, has gained a special place in the pharmaceutical and food industries. The application of these compounds in food matrices promotes the improvement of the organoleptic, functional, and microbiological characteristics of these foods, allowing, among others, the increase in useful life. For these reasons, and especially for the health benefits that can bring the consumption of these improved foods, the demand in the production of these biocompounds has increased.

For the efficient production of biocompounds from agro-industrial waste, it is necessary to scale up the recovery, concentration and purification operations. One of the first steps is the process of separating the biocompounds from the material where they are found. Traditionally, this process is carried out with chemical solvents; however, it has been demonstrated that the improper use of these can cause damage to the environment and to the personnel who handle them. Other alternative processes to traditional extraction, such as maceration or the use of other natural solvents such as water at high temperatures, have been proposed; however, yields are low. Therefore, alternative methodologies have been proposed such as ultrasound, microwaves, use of supercritical fluids, and high voltage electrical discharges (Roselló-Soto et al., 2019; Shahram and Dinani, 2019). Therefore, this chapter will describe the alternative processes for the separation, extraction, and purification of biocompounds.

14.2   RECOVERY PROCESSES

Nowadays, health and environmental problems are raising awareness among the world population about the importance of a balanced diet and about the treatment of waste generated by industries. Researchers in the area of food have a great challenge to reduce and control these problems. In the last years, alternative methodologies for the management of agro-industrial waste have been evaluated; for example, the use of this waste to obtain products of interest such as feed, manufacture of biodegradable products, as a substrate for microbial growth, or for the extraction of metabolites with beneficial effect for health. These compounds are considered bioactive compounds and have a great potential in the treatment and prevention of chronic and acute diseases. Its properties are attributed mainly to the antioxidant activity possessed by the mentioned compounds (Barba and Grimi, 2015).

For example, some citrus residues from food industry have been used to obtain bioactive compounds such as limonene, essential oils (Wu et al., 2017; Golmohammadi et al., 2018; Torres-León et al., 2018), among others. Whey, a by-product of the dairy industry, has been employed as a substrate for the growth of some microorganisms and the subsequent production of bio-compounds (Valduga et al., 2014; Wang et al., 2014). On the other hand, residues such as cane bagasse, molasses, mango, banana, and pomegranate peel and glycerol, have been useful as a substrate for yeast and fungi for the production of enzymes, pigments, proteins, antioxidants, etc. (Schneider et al., 2013; Venil et al., 2013; Almeida Lopes et al., 2017; Mukherjee et al., 2017). Likewise, in the semidesert regions from northern Mexico, some endemic plant materials have been used to obtain metabolites with antioxidant and antimicrobial activities. *Flourensiacernua*, known as tarbush, is a species that grows in the semiarid zones of Mexico. It is a plant rich in polyphenols, benzofuran, lactone, etc. (Castillo et al., 2012; Jasso de Rodríguez et al., 2012).

For the use of all these biocompounds have used different techniques of extrusion in which organic solvents and high temperatures are used (Tellez et al., 2001; Jasso de Rodríguez et al., 2017; Salas-Méndez et al., 2019). Nevertheless, the development of emerging technologies has allowed the application of eco-friendly methodologies, with lower processes costs, shorts operation times and promising results.

14.2.1   CELL RUPTURE

Some microorganisms can produce numerous metabolites of interest such as proteins, enzymes, polysaccharides, pigments, lipids, polyphenols, etc. These compounds are a great opportunity in the prevention of diseases related with oxidative stress, cellular damage, due to the accelerated lifestyle. However, in some cases, the biocompounds production is intracellular type, which represents a disadvantage when it comes to obtaining mentioned metabolites, because it is necessary to disrupt the cell walls and membranes prior to the extraction (Patel et al., 2019). Nevertheless, there are many techniques used for breaking the microbial cell wall. These techniques facilitate the extraction and recovery processes. Lee et al. (2012) propose the classification of the cell rupture techniques in mechanical and nonmechanical. Methods such as bead mill, high pressure, and ultrasonic are considered mechanical methods. Within nonmechanical techniques, we can find osmotic shock, microwave, solvents, enzymatic lysis, supercritical CO_2, etc. The methodology to choose will depend on the producing microorganism, the characteristics of the compounds to be obtained, the ease of scaling the process and the economic parameters (Momin et al., 2018). Also, the use of certain rupture methods depends on cell properties such as physical strength and location of the desired products inside of the cell (Klimek-Ochab et al., 2011; Skorupskaite et al., 2019). For example, to obtain pigments and lipids from plants and some microorganism's species, conventional organic solvent such as chloroform, methanol, ether, hexane, or maceration and soxhlet method have been used. But in the case of microalgae, resulting different difficulties to extraction processes due to the stiffness of cell wall constituted of biomolecules (protein, algaenan, polysaccharides, etc.) with complex linking (Barba and Grimi, 2015; Alhattab et al., 2019; Nagappan et al., 2019). On the other hand, the efficiency of lipid and pigments extraction is dependent on the solvent polarity and mixture of solvent (Byreddy et al., 2015). In the case of protein extraction, it is very important put attention in some factors such as temperature, time of operation, and pH of solution, due the high sensibility of proteins and its possible denaturation. Below, some of the most used methods in cell breakdown and the application they have had, as well as some of the advantage and disadvantage that appear when using them are detailed.

14.2.2   BEAD MILLS

Bead milling have been used to bacteria, microalgae, yeast, and fungal biomass to assist extraction processes (Dong et al., 2016). Also, bead mills have been applied in large-scale processes due to its high disruption efficiency and easy scale-up procedures. The disruption of cell walls by bead mills is carried on by the compression caused for the solid beads that are moved at high velocities (Lee et al., 2017). The efficiency of bead mills depends on different parameters such as biomass concentration, type and diameter of bead, and agitator speed (Lee et al., 2012; Postma et al., 2017). There are some reports about the material of beads (such as glass, zirconia–silica, titanium, zirconium oxide, among others), where it has been found that zirconium oxide beads present a major efficient than glass beads due its specific density (Postma et al., 2017).

14.2.3   ULTRASONICATION

Ultrasonic method has been used for breaking algal, fungi, and yeast cells. It is an efficient, simple, and environment-friendly system (Skorupskaite et al., 2019). During the treatment with ultrasonic system, the energy of high frequency sonic waves causes a cavitation phenome. During cavitation, microbubbles are generated and then, the pressure released when these bubbles collapse, is the responsible of the cell wall disruption, allowing solvent penetration and the solute-solvent contact; therefore, extracts are released (Günerken et al., 2015; Liu et al., 2016; Urnau et al., 2018; Patel et al., 2019). The efficiency of cell disruption by ultrasonication depends of factors such as reaction time, temperature of medium, and viscosity of the cell suspension (Lee et al., 2012). However, Patel et al. (2019), indicate that the extended time for ultrasonication could generate free radicals in solution and this free radical could oxidize lipids or in other case, the biocompounds could be degraded. Also, Gazor et al. (2018) mentioned that ultrasonication method is not suitable for therapeutic proteins because the excessive heat generated could to influence on stability and biological activity of proteins.

14.2.4   HIGH-PRESSURE HOMOGENIZATION

The cell disruption by this methodology is due to the high pressure applied by a piston that rapidly and brusquely breaks the cells when these pass through an adjustable orifice discharge valve (Klimek-Ochab et al., 2011; Safi et al., 2015). It is used at large scale and is an economic and easily scalable method (Dong et al., 2016; Mevada et al., 2019). For example, high pressure homogenizers has been used in dairy industry to reduce fat droplets in milk and for the destruction of microorganisms (Lee et al., 2012; Dong et al., 2016).

14.2.5   OSMOTIC SHOCK

This methodology allows the cell disruption due to the different pressures of solutes. Some studies have been carried on with NaCl and sorbitol (Dong et al., 2016), hypertonic solution with sucrose, EDTA and Tris-HCl (Klimek-Ochab et al., 2011). Byreddy (Byreddy et al., 2015) reported the comparison of cell disruption methods for lipid extraction from *Schizochytrium sp.* S31 and *Thraustochytrium* sp. AMCQS5-5. Some techniques evaluated were bead vertexing, water bath, osmotic shock, sonication, and shake mill. The authors indicate that the maximum lipid extraction yields were obtained with osmotic shock method (29.1 % from *Thraustochytrium* sp. AMCQS5-5 and 48.7% from *Schizochytrium* sp. S31). However, various studies showed that osmotic shock is a very limited method for the application in cells of rigid walls (Dong et al., 2016).

14.2.6   ENZYMATIC METHODS

Enzymatic lysis is a nondestructive treatment and can prevent the degradation of thermally sensitive components (Dong et al., 2016). Also, some enzymes used in the processes can be obtained from biotechnological via, taking advantage of some industrial waste as a substrate for microorganisms producing enzymes. Hernández-Almanza et al. (2017) reported the use of β-1,3-glucanase from *Trichoderma harzianum* for the extraction of carotenoids produced by the red yeast *Rhodotorulaglutinis*. The authors mentioned that enzymatic lysis allows major yields than microwave and ultrasonication methods. However, they are limited to laboratory scale use for bioanalytical purpose duo to their economic and operational limitations (Gazor et al., 2018).

The chemical features and composition of cell walls are important factors for applications in biotechnological processes, also, the easily to recovery the intracellular product is necessary (Lee et al., 2017). Some cell disruption techniques have been developed to establish an low-cost, to optimizing the process, with the efficient use of industrial by-products and effective release of intracellular products (Klimek-Ochab et al., 2011; Almeida-Lopes et al., 2017).

14.3   SEPARATION PROCESSES

After cell rupture processes, it is necessary to apply methodologies to recover the biocompounds. The processes of separation and recovery of the biocompound of the matrix, generally include methods of extraction, whose purpose is to concentrate this biocompound in the medium where it is. The process traditionally used is liquid-liquid or solid-liquid extraction, which allows recovering a solute (biocompound) from a solution using a solvent. The solvent used must be little soluble or insoluble in the solution and must have an affinity with the compound to be recovered (Tejada et al., 2011). However, this methodology has several disadvantages, among which are the pollution generated using large quantities of organic solvents and the low efficiency of the process. Therefore, emerging technologies have been sought that are eco-friendly and improve yields. Among the processes applied are: High voltages electrical discharges (HVED), pulsed electric fields, supercritical fluid, subcritical fluid, pressurized liquid extraction, temperature-induced phase separation, three-phase partitioning, aqueous two-phase extraction, and membrane separation, some of which will be described below.

14.3.1   HIGH VOLTAGES ELECTRICAL DISCHARGES (HVED)

The HVED process is considered an eco-friendly and efficient process, which achieves the extraction of desirable compounds through the use of organic solvents, but in smaller quantities and using less energy. The process uses high electric field between one or multiple point electrode/s and a grounded plate electrode. The basic mechanism of the process is the corona or electric wind that is generated when high voltage is applied between one or more point electrodes and the electrode that is grounded. The ions generated around the point electrodes accelerate and begin to collide with air molecules, creating new charged particles. This impact of electrons on the sample causes the cell wall to rupture and therefore allows the release of the compounds into the medium (Xi et al., 2017; Shahram et al., 2019). Like most technologies, their use has advantages and disadvantages. Among the advantages, it is worth mentioning: (1) efficiency in cell destruction, caused by electric shock, which releases intracellular compounds by diffusion into the environment. Under these conditions the extraction yield is high, the extraction time is short, and there is low energy consumption; (2) the use of solvent is low, the solid-liquid ratio must be low to perform the energy discharge; (3) the working temperature is not high allowing the integrity of the compounds to be extracted to be maintained; (4) some studies have shown that there is an increase in the biological activity of the compounds. Among the disadvantages is that: (1) it is not a selective technique, so it destroys all cells with electric shock; (2) scaling conditions are not yet completely established, so it has not been easy to use at an industrial level; and (3) due to the presence of free radicals in the sample, electric shock could cause damage to the compounds. However, in spite of the disadvantages, it is a methodology that presents interesting performances and its application can solve the problems in the extraction of compounds (Li et al., 2019).

Research carried out by Shahram et al. (2019) using orange pomace to extract phenolic compounds by means of HVED showed that with 10 min of operation the yields were high compared to other studies obtaining a final concentration of 617.76 ± 6.15 mg/L. El Kantar et al. (2019) used electric discharges as pretreatment to extract phenolic compounds with antioxidant activity of grapefruit peel, finding that once applied the method improves up to six times the diffusion of naringin in the medium. Yan et al. (2018) evaluated a continuous system of electric discharges to improve the extraction of flavonoids peanut shells, finding that with the designed system similar yields to the traditional system are reached, but in less time (35 min less), which represents, among others, reduction of process costs. These studies show the feasibility of using this methodology to extract different compounds with interesting biological activities that can be applied to the food industry.

14.3.2   PULSED ELECTRIC FIELDS

Electric pulse fields are another technology that has gained importance in recent years because it is considered eco-friendly. This technology consists in the application of high external electric fields (1–50 kV/cm) in short times to the cell material inducing electroporation, which means an increase in the permeability of the cell membrane to the transport of ions and other molecules. This facilitates the diffusion of biocompounds into the medium, increasing the extraction yield. Some of the advantages of this technology are its low cost, reduces the loss of thermosensitive compounds and does not cause serious damage to the cell (Barba et al., 2016; Redondo et al., 2018). Several studies have been carried out for the extraction of biocompounds with biological activity from agro-industrial by-products. El Kantar et al. (2018) evaluated the effect of the application of electrical pulses in the concentration of phenolic compounds in the shells of orange, grapefruit, and lemon fruits, for later extraction with ethanol mixtures, finding that there were improvements in yields up to 50% without altering the stability of the compounds. Similar studies carried out by Segovia et al. (2015) for the extraction of phenolic compounds from borage leaves, found that pretreatment with electrical pulses increases the performance in the operation, improves the antioxidant activity of compounds and decreases processing times. Sarkis et al. (2015) evaluated the effect of the application of electrical pulses and electric discharges as pretreatments on sesame cake to increase the diffusion of lignans, polyphenols, proteins, and other compounds in the medium for extraction with water and ethanol as solvents. They found that both technologies have application potential allowing the extraction of compounds in less time.

14.3.3   SUPERCRITICAL FLUID

One of the technologies that has gained importance in recent decades, classified as "green technology" is extraction using supercritical fluids. A supercritical fluid is any substance that is at a temperature and pressure above its thermodynamic critical point. Among the most commonly used fluids is CO_2, as it has low or no toxicity, is not flammable corrosive, colorless, economical, easy disposal without leaving residues and one of the most important, it is used at low temperatures. Supercritical fluids have the property of diffusing through solids as a gas but dissolving compounds as a liquid. They can easily change their density with temperature or pressure. Therefore, they are alternative fluids to organic solvents in compound extraction processes (Velasco et al., 2007; Barba et al., 2016). Roselló-Soto et al. (2019) evaluated the use of the supercritical fluid CO_2 as an alternative for the extraction and recovery of oils from by-products of "horchata", finding that the content of α-tocopherol was twice as high in the samples treated by this methodology, as well as increased the content of phenolic compounds and total antioxidant activity, so it is suggested that this could be an alternative to the conventional extraction process.

14.3.4   AQUEOUS TWO-PHASE SYSTEM

One of the separation technologies that functions not as pretreatment for extraction but as an integral methodology (extraction and separation) is the technology is the two-phase aqueous system. This is a liquid-liquid extraction technique that allows the separation of biocompounds after having applied a biotechnological process. It has some advantages such as rapid separation, extraction performance, compatibility with various compounds among others. The system consists of the mixture in an aqueous medium of two different polymers of flexible chain (ethylene oxide propylene copolymer of oxide (EOPO), polyethylene glycol (PEG), polyacrylates, dextran) or a combination of polymer of low molecular weight, and a salt above a critical concentration (phosphate, citrate, and sulphate), where two phases are generated, one rich in polymer and the other rich in salt. According to the affinity of the biocompound of interest, its molecular weight, surface load and hydrophobicity, and the conditions of the system, this compound will remain in one of the two phases and will be separated. This is a selective technique and in recent years its study has increased (Benavides and Rito-Palomares, 2008; Bravo et al., 2011; Iqbal et al., 2016; Phong et al., 2018). Using biphasic systems, proteins (Wu et al., 2017), enzymes (Rahimpour et al., 2016; Shad et al., 2018), and other compounds of interest for application in the food industry have been recovered.

14.4 PURIFICATION PROCESSES

Purification processes are used after separation processes in order to obtain more purified substances, and to be able to determine their biochemical properties and their application in industry (Yepes et al., 2008). The most commonly used purification processes today are precipitation (Bhati et al., 2013; Sharma et al., 2017), ultrafiltration and/or aqueous biphasic systems, complemented by final techniques involving multiple chromatographic stages, using ion exchange columns and gel filtration (Brandelli et al., 2015). Table 14.1 shows the main advantages and disadvantages of the purification processes used in industry.

TABLE 14.1 Advantages and Disadvantages of Different Purification Processes.

Purification processes	Advantages	Disadvantages	References
Precipitation with organic solvents or salts	It is a simple, economical, easy to scale, widely used technique.	The use of organic solvents, denatures enzymes, requires a step of dialysis (desalination) after the precipitation of salt, generally of low resolution.	(Habbeche et al., 2014)

TABLE 14.1 *(Continued)*

Purification processes	Advantages	Disadvantages	References
Two-phase aqueous system	Low cost, good resolution, scalable, able to remove cellular waste, does not affect proteins, slightly toxic.	Determination of proteins can be affected by polymers and salts, usually changes in phase viscosity occur.	Sala et al., 2014
Ultrafiltration	Minimal denaturation of enzymes by low pressures and temperatures, simple technique, protein concentration, easy to scale	Expensive membrane technique, low resolution, flow permeation can be reduced by membrane fouling and polarization.	Rajput and Gupta, 2013
Chromatography with ion exchange column	High resolution, decreases sample volume, easy to scale, protein binding capacity increases.	Costly process, lower throughput when compared to non-chromatographic techniques.	Poopathi et al., 2014
Chromatography with gel filtration column	Desalination, high resolution.	Difficult to scale, low performance compared to other non-chromatographic methods, expensive process, diluted sample at the end of the process.	Paul et al., 2014

Purification processes are generally used after organic substance extraction processes (Saini et al., 2018) such as pectin, enzymes, essential oils, flavonoids, carotenes, antioxidants, amino acids, dietary fiber, vitamins, and colorants (Chagas et al., 2017), compounds found in waste or by-products of agro-industrial processes. In recent years, research has focused on the use of agro-industrial by-products, which are not used and are directly discarded (Tacin et al., 2019). Table 14.2 shows purification processes for different agro-industrial by-products.

TABLE 14.2   Purification Processes for Different Agro-Industrial By-Products.

Agro-industrial by-product	Organic substance extracted	Purification method	Use	Reference
Passion fruit peel	Pectin	Precipitation	Gelling, stabilizing, and thickening agent in food systems such as jams and jellies, confectionery and fruit juice, pharmaceutical activities, including wound healing, lipase inhibition, apoptosis induction of human cancer cell, as well as immune stimulant, antimetastatic, anti-ulcer, and cholesterol decreasing effects.	Freitas de Oliviera et al., 2016
Oils from agro-industrial wastes	Hydrolases enzymes: Lipases	Precipitation	Detergents, cosmetics, water treatment, bioremediation, oleochemical	Tacin et al., 2019
Red and white grape pomace containing seeds and skins	Polyphenols	Ultrafiltration: membrane 0.45 nm	Development of new nutraceuticals, functional foods, additives, pharmaceutical industry, and in medicine.	Martins et al., 2016
Marine shells	Proteases	Precipitation using ammonium sulphate (75% saturation), dialyzed against Tris-HCl buffer (pH 7.2; 50 mM) chromatography with gel filtration column	Industrial sectors such as detergent, food, pharmaceutical, chemical, leather and silk, apart from waste treatment.	Maruthiah et al., 2015
Seeds, tomato pulp	Peptides	Ultrafiltration and chromatography with gel filtration column packed with Sephadex G120 25 gel.	Depending on the composition and sequence of amino acids, bioactive peptides of 50 lengths may show different activities, such as antihypertensive, antioxidant, immunomodulator, opioid, anti-inflammatory, or antimicrobial.	Moayedi et al., 2018

Precipitation is a purification technique most used for the extraction of pectin and enzymes from agro-industrial waste (Martins et al., 2016), organic salts, and solvents can be used to generate precipitation (Brandelli et al., 2015). Generally, copper sulphate, ammonium sulphate, and acetone are used after dialysis. This purification process is almost always complemented by other purification processes such as chromatography with gel filtration column in Sephadex G-200, G-100, Sephacryl S-200 (Correa et al., 2010), and ion exchange column using a matrix of Q-Sepharose, SP-Sepharose, and DEAE-Sepharose (Cavello et al., 2013), which seek a high purification. But there is a reduction of the purified volume, while if the precipitation is combined with non-chromatographic techniques such as two-phase aqueous systems a large volume of the purified is obtained (Bach et al., 2012), aqueous systems are formed by PEG and salts. The presence of PEG can interfere in some analytical determinations, such as electrophoresis and determination of proteins by the Lowry method, for this reason the aqueous systems are combined with the application of ultrafiltration in diafiltration mode to remove PEG (Sala et al., 2014). The choice of purification method will depend on the purpose desired.

14.5   PERSPECTIVES

In recent years, the interest in using biocompounds in the development of functional foods has increased but one of the limitations continues to be the processes carried out to obtain these compounds. The methods of separation and purification of compounds using alternative technologies to the traditional, environment-friendly continue to be the subject of research. It is necessary to approach from different angles to improve the performance of the processes, increase the range of biocompounds that can be separated, reduce the use of toxic organic solvents or possible residues considering the tendencies of circular economy, reduce processing times, energy consumed and of course, consider scaling to make its application feasible at an industrial level. Likewise, it is necessary to study the selectivity of the processes to reduce or integrate the unitary operations carried out, allowing to carry out operations of cellular rupture, extraction of compounds, and purification in less stages. The knowledge of these processes can potentiate the use of green technologies in the extraction and purification of biocompounds with biological activities from agro-industrial waste.

KEYWORDS

- **biocompounds**
- **cell disruption**
- **recovery processes**
- **purification methods**
- **emerging technologies**

REFERENCES

Alhattab, M.; Kermanshahi-Pour, A.; Brooks, M. S. Microalgae Disruption Techniques for Product Recovery : Influence of Cell Wall Composition. *J. Appl. Phycol.* **2019,** *1*, 61–88.

Almeida Lopes, N.; Remendi, R.; dos Santos C.; Veiga, C. A.; Fernandes, J. Different Cell Disruption Methods for Obtaining Carotenoids by Sporodiobolus pararoseus and Rhodothorula mucilaginosa. *Food Sci. Biotechnol.* **2017,** *26*(3), 759–766.

Bach, E.; Sant´Anna, V.; Daroit, D. J.; Correa, A, P, F., Segalin, A.; Brandelli, A. Production, One-Step Purification, and Characterization of a Keratinolytic Protease from Serratia marcescens P3. *Process Biochem.* **2012,** *47*, 2455–2462.

Barba, F. J.; Grimi, N. New Approaches for the Use of Non-conventional Cell Disruption Technologies to Extract Potential Food Additives and Nutraceuticals from Microalgae. *Food Eng. Rev.* **2015,** *7*, 45–62.

Barba, F. J.; Zhu, Z.; Koubaa, M.; Sant'Ana, A. S.; Orlien, V. Green Alternative Methods for the Extraction of Antioxidant Bioactive Compounds from Winery Wastes and By-Products: A Review. *Trends Food Sci. Technol.* **2016,** *49*, 96–109.

Benavides, J.; Rito-Palomares, M. Aplicación genérica de sistemas de dos fases acuosas Polietilénglicol - Sal para el desarrollo de procesos de recuperación primaria de compuestos biológicos. *Rev. Mex. Ing. Química.* **2008,** *7*. http://www.scielo.org.mx/pdf/rmiq/v7n2/v7n2a2.pdf

Bhati, M.; Khokhar, D.; Pandey, A.; Gaur, A. Purification and Characterization of Lipase from Aspergillus japonicas: A Potent Enzyme for Biodiesel Production. *Natl. Acad. Sci. Lett.* **2013,** *36*(2), 151–156.

Brandelli, A.; Sala, L.; Kalil, S. J. Microbial Enzymes for Bioconversion of Poultry Waste into Added-Value Products. *Food Res. Int.* **2015,** *73*, 3–12.

Bravo, K. E.; Muñoz, K.: Calderon, J.; Osorio, E. J. Desarrollo de un método para la extracción de Polifenol Oxidasa de Uchuva (Physalis peruviana L.) y aislamiento por sistemas bifásicos acuosos. *Vitae.* **2011,** *18*(2), 124–132. https://www.redalyc.org/html/1698/169822670003/

Byreddy, A. R.; Gupta, A.; Barrow, C. J.; Puri, M. Comparison of Cell Disruption Methods for Improving Lipid Extraction from Thraustochytrid Strains. *Mar. Drugs.* **2015,** *13*(8), 5111–5127.

Castillo, F.; Hernández, D.; Gallegos, G.; Rodríguez, R.; Aguilar, C. N. Antifungal Properties of Bioactive Compounds from Plants. In *Fungicides for Plant and Animal Diseases*; Dhanasekaran, D., Ed.; 2012.

Cavello, I. A.; Hours, R. A.; Rojas, N. L.; Cavalito, S. F. Purification and Characterization of a Keratinolytic Serine Protease from Purpureocillium lilacinum LPS # 876. *Process Biochem.* **2013**, *48*, 972–978.

Chagas, R.; Santana, J.; Denadai, M.; Nunes, M. Evaluation of Bioactive Compounds Potential and Antioxidant Activity in some Brazilian Exotic Fruit Residues. *Food Res. Int.* **2017**, *102*, 84–92.

Correa, A.; Daroit, D.; Brandelli, A. Caracterización de una queratinasa producida por Bacillus sp. P7 aislado de un entorno amazónico. *Int. Biodeterior. Biodegradation.* **2010**, *64*, 1–6.

Dong, T.; Knoshaug, E. P.; Pienkos, P. T.; Laurens, L. M. L. Lipid Recovery from Wet Oleaginous Microbial Biomass for Biofuel Production : A Critical Review. *Appl. Energy.* **2016**, *177*, 879–895.

El Kantar, S.; Boussetta, N.; Lebovka, N.; Foucart, F.; Rajha, H. N.; Maroun, R. G.; Louka, N.; Vorobiev, E. Pulsed Electric Field Treatment of Citrus Fruits: Improvement of Juice and Polyphenols Extraction. *Innov. Food Sci. Emerg. Technol.* **2018**, *46*, 153–161.

El Kantar, S.; Rajha, H. N.; Boussetta, N.; Vorobiev, E.; Maroun, R. G.; Louka, N. Green Extraction of Polyphenols from Grapefruit Peels using High Voltage Electrical Discharges, Deep Eutectic Solvents and Aqueous Glycerol. *Food Chem.* **2019**, *295*, 165–171.

Freitas de Oliviera, C.; Giordani, D.; Lutckemier, R.; Deyse Gurak, P.; Cladera Olivera, F.; Ferreira Marczack, L. D. Extraction of Pectin from Passion Fruit Peel Assisted by Ultrasound. *Food Sci. Technol.* **2016**, *71*, 110–115.

Gazor, M.; Siamak, S.; Talesh, A.; Khatami, M.; Javidanbardan, A. A. Novel Cell Disruption Approach : Effectiveness of Laser-induced Cell Lysis of Pichia pastoris in the Continuous System. *Biotechnol. Bioprocess Eng.* **2018**, *23*, 49–54.

Golmohammadi, M.; Borghei, A.; Zenouzi, A.; Ashrafi, N.; Taherzadeh, M. J. Optimization of Essential Oil Extraction from Orange Peels Using Steam Explosion. *Heliyon* **2018**, *4*(11), e00893.

Günerken, E.; Hondt, E. D.; Eppink, M. H. M.; Garcia-gonzalez, L.; Elst, K.; Wijffels, R. H. Cell Disruption for microalgae biore fi neries. **2015**, *33*, 243–260.

Habbeche, A.; Saudi, B.; Jaouadi, B.; Haberra, S.; Kerouaz, B.; Boudelaa, M. Purification and Biochemical Characterization of a Detergent-Stable Keratinase from a Newly Thermophilic Actinomycete Actinomadura keratinilytica strain Cpt29 Isolated from Poultry Compost. *J. Biosci. Bioeng.* **2014**, *117*, 413–421.

Hernández-Almanza, A.; Navarro-Macías, V.; Aguilar, O.; Aguilar-González, M.; Aguilar, C. N. Carotenoids Extraction from Rhodotorula glutinis Cells Using Various Techniques : A Comparative Study. *Indian J. Exp. Biol.* **2017**, *55*, 479–484.

Iqbal, M.; Tao, Y.; Xie, S.; Zhu, Y.; Chen, D.; Wang, X.; Huang, L.; Peng, D.; Sattar, A.; Shabbir, M. A. B.; Hussain, H. I.; Ahmed, S.; Yuan, Z. Aqueous Two-Phase System (ATPS): An Overview and Advances in its Applications. *Biol. Proced.* **2016**, *18*(1), 1–18.

Jasso de Rodríguez, D.; Hernández-Castillo, F. D.; Solís-Gaona, S.; Rodríguez-García, R.; Rodríguez-Jasso, R. M. Flourensia cernua DC: A Plant from Mexican Semiarid Regions with a Broad Spectrum of Action for Disease Control. In *Integrated Pest Management and Pest Control - Current and Future Tactics*; Soloneski, S., Ed.; 2012.

Jasso de Rodríguez, D.; Salas-Méndez, E. J.; Rodríguez-García, R.; Hernández-Castillo, F. D.; Díaz-Jiménez, M. L. V.; Sáenz-Galindo, A.; González-Morales, S.; Flores-López, M. L.; Villarreal-Quintanilla, J. A.; Peña-Ramos, F. M.; Carrillo-Lomelí, D. A. Antifungal Activity in vitro of Ethanol and Aqueous Extracts of Leaves and Branches of Flourensia spp. Against Postharvest Fungi. *Ind. Crops Prod.* **2017**, *107*, 499–508.

Klimek-Ochab, M.; Brzezinska-Rodak, M.; Zymanczyk-Duda, E.; Lejczak, B.; Kafarski, P. Comparative Study of Fungal Cell Disruption—Scope and Limitations of the Methods. *Folia Microbiol. (Praha).* **2011**, *56*, 469–475.

Lee, A. K.; Lewis, D. M.; Ashman, P. J. Disruption of Microalgal Cells for the Extraction of Lipids for Biofuels: Processes and Specific Energy Requirements. *Biomass Bioenerg.* **2012**, *46*, 89–101.

Lee, S. Y.; Cho, J. M.; Chang, Y. K.; Oh, Y. K. Cell Disruption and Lipid Extraction for Microalgal Biorefineries: A Review. *Bioresour. Technol.* **2017**, *244*, 1317–1328.

Li, Z.; Fan, Y.; Xi, J. Recent Advances in High Voltage Electric Discharge Extraction of Bioactive Ingredients from Plant Materials. *Food Chem.* **2019**, *277*, 246–260.

Liu, D.; Ding, L.; Sun, J.; Boussetta, N.; Vorobiev, E. Yeast Cell Disruption Strategies for Recovery of Intracellular Bio-Active Compounds—a Review. *Innov. Food Sci. Emerg. Technol.* **2016**, *36*, 181–192.

Martins, I. M.; Bruna, S. R.; Blumberg, J.; Chen, C. Y. O.; Macedo, G. Enzymatic Biotransformation of Polyphenolics Increases Antioxidant Activity of Red and White Grape Pomace. *Food Res. Int.* **2016**, *89*, 533–539.

Maruthiah, T.; Somanath, B.; Immanuel, G.; Palavesam, A. Deproteinization Potential and Antioxidant Property of Haloalkalophilic Organic Solvent Tolerant Protease from Marine Bacillus sp. APCMST-RS3 Using Marine Shell wastes. *Biotechnol. Reports* **2015**, *8*, 124–132.

Melgarejo-Moreno, J. Agua y economía circular. Congr. Nac. del Agua Orihuela **2019**, 27–52. https://rua.ua.es/dspace/bitstream/10045/88467/1/Congreso_Nacional_Agua_2019_27-52.pdf

Mevada, J.; Devi, S.; Pandit, A. Large scale Microbial Cell Disruption Using Hydrodynamic Cavitation : Energy Saving Options. *Biochem. Eng. J.* **2019**, *143*, 151–160.

Moayedi, A.; Mora, L.; Aristoy, M, C.; Safari, M.; Hashemi, M.; Toldrá, F. Peptidomic Analysis of Antioxidant and ACE-Inhibitory Peptides Obtained from Tomato Waste Proteins Fermented Using Bacillus subtilis. *Food Chem.* **2018**, *1*, 180–187.

Momin, B.; Chakraborty, S.; Annapure, U. Investigation of the Cell Disruption Methods for Maximizing the Extraction of Arginase from Mutant Bacillus licheniformis (M09) Using Statistical Approach. *Korean J. Chem. Eng.* **2018**, *35*(10), 2024–2035.

Mukherjee, G.; Mishra, T; Deshmukh, S. K. Fungal Pigments: An Overview. In *Developments in Fungal Biology and Applied Mycology*; Satyanarayana, T., ed.; Springer Nature Singapore: Singapore, 2017, pp 525–541.

Nagappan, S.; Devendran, S.; Tsai, P.; Dinakaran, S. Passive Cell Disruption Lipid Extraction Methods of Microalgae for Biofuel Production – A review. *Fuel* **2019**, *252*(100), 699–709.

Patel, A.; Arora, N.; Pruthi, V.; Pruthi, P. A. A Novel Rapid Ultrasonication-Microwave Treatment for Total Lipid Extraction from Wet Oleaginous Yeast Biomass for Sustainable Biodiesel Production. *Ultrason. Sonochem.* **2019**, *51*, 504–516.

Paul, T.; Das, A.; Mandal, A.; Halder, S.; Das-Mohapatra, P.; Pati, B. Production and Purification of Keratinase Using Chicken Feather Bioconversion by a Newly Isolated Aspergillus fumigatus TKF1: Detection of Valuable Metabolites. *Biomass Convers. Biorefinery* **2014**, *4*, 137–148.

Phong, W. N.; Show, P. L.; Chow, Y. H.; Ling, T. C. Recovery of Biotechnological Products Using Aqueous Two Phase Systems. *J. Biosci. Bioeng.* **2018**, *126*(3), 273–281.

Poopathi, S.; Thirugnanasambantham, K.; Mani, K.; Lakshmi, P, V.; Ragul, K. Purification and Characterization of Keratinase from Feather Degrading Bacterium Useful for Mosquito Control—a New Report. *Trop. Biomed.* **2014**, *31*, 97–109.

Postma, P. R.; Suarez-Garcia, E.; Safi, C.; Olivieri, G.; Olivieri, G.; Wijffels, R. H.; Wijffels, R. H. Energy Efficient Bead Milling of Microalgae: Effect of Bead Size on Disintegration and Release of Proteins and Carbohydrates. *Bioresour. Technol.* **2017**, *224*, 670–679.

Rahimpour, F.; Hatti-Kaul, R.; Mamo, G. Response Surface Methodology and Artificial Neural Network Modelling of an Aqueous Two-Phase System for Purification of a Recombinant Alkaline Active Xylanase. *Process Biochem.* **2016**, *51*(3), 452–462.

Rajput, R.; Gupta, R. Thermostable Keratinase from Bacillus pumilus KS12: Production, Chitin Crosslinking and Degradation of Sup35NM Aggregates. *Bioresour. Technol.* **2013**, *133*, 118–126.

Redondo, D.; Venturini, M. E.; Luengo, E.; Raso, J.; Arias, E. Pulsed Electric Fields as a Green Technology for the Extraction of Bioactive Compounds from Thinned Peach By-Products. *Innov. Food Sci. Emerg. Technol.* **2018**, *45*, 335–343.

Roselló-Soto, E.; Barba, F. J.; Lorenzo, J. M.; Dominguez, R.; Pateiro, M.; Mañes, J.; Moltó, J. C. Evaluating the Impact of Supercritical-CO 2 Pressure on the Recovery and Quality of Oil from "Horchata" By-Products: Fatty Acid Profile, α-Tocopherol, Phenolic Compounds, and Lipid Oxidation Parameters. *Food Res. Int.* **2019**, *120*, 888–894.

Roselló-Soto, E.; Koubaa, M.; Moubarik, A.; Lopes, R. P.; Saraiva, J. A., Boussetta, N.; Grimi, N.; Barba, F. J. Emerging Opportunities for the Effective Valorization of Wastes and By-Products Generated During Olive Oil Production Process: Non-Conventional Methods for the Recovery of High-Added Value Compounds. *Trends Food Sci. Technol.* **2015**, *45*(2), 296–310.

Sadh, P. K.; Duhan, S.; Duhan, J. S. Agro-Industrial Wastes and their Utilization Using Solid State Fermentation: a Review. *Bioresour. Bioprocess.* **2018**, *5*(1), 1–15.

Safi, C.; Frances, C.; Ursu, A. V.; Laroche, C.; Pouzet, C.; Vaca-Garcia, C.; Pontalier, P. Y. Understanding the Effect of Cell Disruption Methods on the Diffusion of chlorella vulgaris Proteins and Pigments in the Aqueous Phase. *Algal Res.* **2015**, *8*, 61–68.

Saini, R.; Moon, S.; Keum, Y. An Updated Review on Use of Tomato Pomace and Crustacean Processing Waste to Recover Commercially Vital Carotenoids. *Food Res. Int.* **2018**, *108*, 516–529.

Sala, L.; Gautérico, G. V.; Younan, F. F.; Brandelli, A.; Morales, C. C.; Kalil, S. J. Integration of Ultrafiltration into an Aqueous two-Phase System in the Keratinase Purification. *Process Biochem.* **2014a**, *49*, 2016–2024.

Sala, L.; Gautério, G. V.; Younan, F. F.; Brandelli, A.; Morales, C. C.; Kalil, S. J. Integration of Ultrafiltration into an Aqueous Two-Phase System in the Keratinase Purification. *Process Biochem.* **2014b**, *49*, 2016–2024.

Salas-Méndez, E. de J.; Vicente, A.; Pinheiro, A. C.; Ballesteros, L. F.; Silva, P.; Rodríguez-García, R.; Hernández-Castillo, F. D.; Díaz-Jiménez, M. de L. V.; Flores-López, M. L.; Villarreal-Quintanilla, J. Á.; Peña-Ramos, F. M.; Carrillo-Lomelí, D. A.; Jasso de Rodríguez, D. Application of Edible Nanolaminate Coatings with Antimicrobial Extract of Flourensia cernua to Extend the Shelf-Life of Tomato (Solanum lycopersicum L.) fruit. *Postharvest Biol. Technol.* **2019**, *150*, 19–27.

Sarkis, J. R.; Boussetta, N.; Tessaro, I. C.; Marczak, L. D. F.; Vorobiev, E. Application of Pulsed Electric Fields and High Voltage Electrical Discharges for Oil Extraction from Sesame Seeds. *J. Food Eng.* **2015**, *153*, 20–27.

Schneider, T.; Graeff-Hönninger, S.; French, W. T.; Hernandez, R.; Merkt, N.; Claupein, W.; Hetrick, M.; Pham, P. Lipid and Carotenoid Production by Oleaginous Red Yeast Rhodotorula glutinis Cultivated on Brewery Effluents. *Energy* **2013**, *61*, 34–43.

Segovia, F. J.; Luengo, E.; Corral-Pérez, J. J.; Raso, J.; Almajano, M. P. Improvements in the Aqueous Extraction of Polyphenols from Borage (Borago officinalis L.) Leaves by Pulsed Electric Fields: Pulsed Electric Fields (PEF) Applications. *Ind. Crops Prod.* **2015**, *65*, 390–396.

Shad, Z.; Mirhosseini, H.; Hussin, A. S. M.; Forghani, B.; Motshakeri, M.; Manap, M. Y. A. Aqueous two-Phase Purification of α-Amylase from White Pitaya (Hylocereus undatus) Peel in Polyethylene Glycol/Citrate System: Optimization by Response Surface Methodology. *Biocatal. Agric. Biotechnol.* **2018**, *14*, 305–313.

Shahram, H.; Taghian Dinani, S. Influences of Electrohydrodynamic Time and Voltage on Extraction of Phenolic Compounds from Orange Pomace. *LWT.* **2019**, *111*, 23–30.

Sharma, P.; Sharma, N.; Pathania, S.; Handa, S. Purification and Characterization of Lipase by Bacillus methylotrophicus PS3 Under Submerged Fermentation and its Application in Detergent Industry. *J. Genet. Eng. Biotechnol.* **2017**, *15*(2), 369–377.

Skorupskaite, V.; Makareviciene, V.; Sendzikiene, E.; Gumbyte, M. Microalgae Chlorella sp. Cell Disruption Efficiency Utilising Ultrasonication and Ultrahomogenisation Methods. *J. Appl. Phycol.* **2019**, *31*(4).

Tacin, M. V.; Massi, F. P.; Fungaro, M. H. P.; Teixeira, M. F. S.; de Paula, A. V.; de Carvalho Santos-Ebinuma, V. Biotechnological Valorization of Oils from Agro-Industrial Wastes to Produce Lipase Using Aspergillus sp. from Amazon. *Biocatal. Agric. Biotechnol.* **2019**, *17*, 369–378.

Tejada, A.; Montesinos, R.; Guzmás, R. *Bioseparaciones.* (Segunda Edicion); Pearson Educación: México, 2011.

Tellez, M.; Estell, R.; Fredrickson, E.; Wedge, D. Extracts of Flourensia cernua (L): Volatile Constituents and Antifungal, Antialgal, and Antitermite Bioactivities. *J. Chem. Ecol.* **2001**.

Torres-León, C.; Ramírez-Guzmán, N.; Londoño-Hernández, L.; Martínez-Medina, G.E. Food Waste and Byproducts: An Opportunity to Minimize Malnutrition and Hunger in Developing Countries. *Front. Sustain. Food Syst.* **2018**, *2*, 1–17.

Urnau, L.; Colet, R.; Soares, V. F.; Franceschi, E.; Valduga, E.; Steffens, C. Extraction of Carotenoids from Xanthophyllomyces dendrorhous using Ultrasound-Assisted and Chemical Cell Disruption Methods. *Can. J. Chem. Eng.* **2018**, *96*, 1377–1381.

Valduga, E.; Rausch Ribeiro, A. H.; Cence, K.; Colet, R.; Tiggemann, L.; Zeni, J.; Toniazzo, G. Carotenoids Production from a Newly Isolated Sporidiobolus pararoseus Strain using Agroindustrial Substrates. *Biocatal. Agric. Biotechnol.* **2014**, *3*(2), 207–213.

Velasco, R. J.; Villada, H. S.; Carrera, J. E. Aplicaciones de los Fluidos Supercríticos en la Agroindustria Applications of Supercritical Fluids in the Agroindustry. *Inf. Tecnológica.* **2007**, *18*(1), 53–65.

Venil, C. K.; Zakaria, Z. A.; Ahmad, W. A. Bacterial Pigments and their Applications. *Process Biochem.* **2013**, *48*(7), 1065–1079.

Wang, Q.; Feng, L.; Luo, W.; Li, H.; Zhou, Y.; Yu, X. Effect of Inoculation Process on Lycopene Production by Blakeslea trispora in a Stirred-Tank Reactor. *Appl. Biochem. Biotechnol.* **2014**, *175*(2), 770–779.

Wu, F.; Jin, Y.; Xu, X.; Yang, N. Electrofluidic Pretreatment for Enhancing Essential Oil Extraction from Citrus Fruit Peel Waste. *J. Clean. Prod.* **2017a**, *159*, 85–94.

Wu, Z.; Hu, G.; Wang, K.; Zaslavsky, B. Y.; Kurgan, L.; Uversky, V. N. What are the Structural Features that Drive Partitioning of Proteins in Aqueous Two-Phase Systems? *Biochim. Biophys. Acta Proteins Proteom.* **2017b,** *1865*(1), 113–120.

Xi, J.; He, L.; Yan, L. Continuous Extraction of Phenolic Compounds from Pomegranate Peel using High Voltage Electrical Discharge. *Food Chem.* **2017,** *230*, 354–361.

Yan, L. G.; Deng, Y.; Ju, T.; Wu, K. Xi, J. Continuous High Voltage Electrical Discharge Extraction of Flavonoids from Peanut Shells Based on "Annular Gap Type" Treatment Chamber. *Food Chem.* **2018,** *256*, 350–357.

Yepes, S. M.; Naranjo, L. J. M.; Sánchez, F. O. Valorizacion de residuos agroindustriales en el Valle de Aburra. *Rev. Fac. Nac. Agroind. Medellin.* **2008,** *61*(1), 4422–4431.

Yusuf, M. Agro-Industrial Waste Materials and their Recycled Value-Added Applications. *Handb. Ecomater.* **2017,** 1–11.

Index

Printed and bound by CPI Group (UK) Ltd, Croydon, CR0 4YY

23/10/2024

01777672-0011